Molecular Biology Biochemistry and Biophysics

26

Arthur S. Brill

Transition Metals in Biochemistry

With 49 Figures

Springer-Verlag Berlin Heidelberg New York 1977

Professor ARTHUR S. BRILL
Department of Physics
University of Virginia
McCormick Road
Charlottesville, VA 22901/USA

ISBN-13: 978-3-642-81150-0 e-ISBN-13: 978-3-642-81148-7
DOI: 10.1007/ 978-3-642-81148-7

Library of Congress Cataloging in Publication Data. Brill, Arthur S. 1927—. Transition metals in biochemistry. (Molecular biology, biochemistry, and biophysics; 26). Bibliography: p. Includes index. 1. Biological chemistry. 2. Transition metals. I. Title. II. Series. QP532.B74. 574.1'921. 77-2787

2131/3130-543210

Preface

Transition metal ions in biological systems are of interest in biology, biochemistry, chemistry, medicine, and physics. Scientists with rather different viewpoints, employing many methods, have contributed to this area. A concise review of the current state of the field will, to some extent, reflect the special knowledge of the person writing it – in this case application of physical methods to the investigation of metal coordination. X-ray diffraction is one of the most important of these methods, but a useful treatment of X-ray structure analysis would be comparable in size with and beyond the scope of the monograph. Many results of X-ray diffraction studies are, of course, presented. Electron paramagnetic resonance spectroscopy has played a major part in the rapid advance in knowledge of the electronic structures of transition metal ions in biological systems. More generally, measurements involving light, microwaves, and magnetic fields are capable of producing much new information, and the required instrumentation is available at most research institutions. Therefore light absorption and paramagnetic resonance are treated in depth. The principles described in the latter discussions are broadly applicable, for example to the promising techniques of X-ray spectroscopy (utilizing synchrotron radiation) and lanthanide-perturbed, very high-resolution nuclear magnetic resonance spectroscopy.

Those aspects of theory are presented which can be directly employed in the quantification and interpretation of experimental data; sufficient background, formulas, and examples are given so that the reader is equipped to understand similar analyses in the literature and to begin to proceed on his own. Only material generally accepted as factual is given in the text; references are provided so that the reader is able to find and evaluate for himself recent data and current speculation. The author believes that the monograph can aid in providing a basis for closer communication among scientists of different backgrounds with a common interest in the biochemistry of transition metal ions.

I thank many colleagues for help in the forms of photographs of figures, references, criticism, and encouragement.

Charlottesville, Virginia
April, 1977

ARTHUR S. BRILL

Contents

Chapter 1

The Role of Transition Metal Ions in Biological Oxidation and Related Processes

Proteins perform many functions in support of life. Some serve as structural materials; others transport gases in the bloodstream between lungs and tissues; still others are involved in blood coagulation; a few are hormones. Large in number and diverse in structure are the enzymes, catalysts for the multitude of chemical reactions of degradation, synthesis, and transformation continuously going on in all biological systems. Many enzymes, particularly those involved in oxidation-reduction reactions, contain metal ions of the first and second "transition series" as "prosthetic groups". We begin by describing what is meant by the terms "transition series" and "prosthetic groups".

1. Transition Metal Ions

The extra-nuclear description of an atom or ion starts with its ground state configuration which specifies the number of electrons in each spatial orbital. Thus the electron configuration of argon is

$$(1s)^2(2s)^2(2p)^6(3s)^2(3p)^6$$

where, for example, $2p^6$ indicates that there is a pair of electrons in each of the three spatial orbitals characterized by principal quantum number 2 and orbital angular momentum quantum number 1 (see Chap. 6-1). The 2p "shell" can, by the Pauli principle, hold just six electrons and is seen to be filled in this example. Indeed, all s, p, and d shells can contain at most two, six, and ten electrons respectively.

The transition groups or series in the periodic table are those in which an inner shell is being filled with electrons while an outer shell of slightly lower energy remains populated. In the first or iron group of transition elements, it is the 3d shell which receives additional electrons after the 4s shell has initially received two. The latter two electrons come into 4s with potassium and calcium, and the first transition element is scandium with the following electron configuration:

$$(1s)^2(2s)^2(2p)^6(3s)^2(3p)^6(3d)^1(4s)^2.$$

The ground configurations of ions of the iron group are given in Table 1-1. While zinc is not considered a transition element, we include the zinc ion in Table 1-1 so that it can be related electronically to members of the 3d group. Table 1-1 is discussed

Table 1-1. Ground states of ions of the iron group

Ions	Ground configuration[a]	Free ion ground state
Ti(III)	$3d^1$	^2D
V(III)	$3d^2$	^3F
Cr(III),V(II)	$3d^3$	^4F
Cr(II)	$3d^4$	^5D
Fe(III),Mn(II)	$3d^5$	^6S
Co(III),Fe(II)	$3d^6$	^5D
Co(II)	$3d^7$	^4F
Ni(II)	$3d^8$	^3F
Cu(II)	$3d^9$	^2D
Zn(II),Cu(I)	$3d^{10}$	^1S

[a]Not explicitly shown is the closed shell (argon) inner configuration, e.g., Ti(III): $1s^2\ 2s^2\ 2p^6\ 3s^2\ 3p^6\ 3d^1$

further in Chapter 6-3. The iron series has among its members several elements of widespread biological occurrence and of great importance in oxygen transport and utilization, and, more generally, in biological oxidation-reduction catalysis, biosynthesis, and in other metabolic processes. The next transition series, the palladium group, is characterized by the filling of the 4d shell. In this series only molybdenum plays an appreciable role in living systems. The remaining transition groups (the lanthanides or "rare earths", a 4f series; the platinum group, a 5d series; and the actinide series, 5f) are not known to be of biological importance.

2. Prosthetic Groups

As an adjective, the word "protein" describes molecules, or those parts of molecules, which are composed entirely of α-amino acid residues, covalently united head-to-tail by peptide bonds to form long, unbranched polymers. These polymers yield only amino acids upon hydrolysis. The reader is assumed to be familiar with the structural chemistry of such polypeptide chains. As a noun, the word protein is commonly used for two kinds of molecule. "Simple" proteins are as just described. In "conjugated" proteins, there are, in addition, firmly bonded non-amino acid com-

Fig. 1-1. Heme

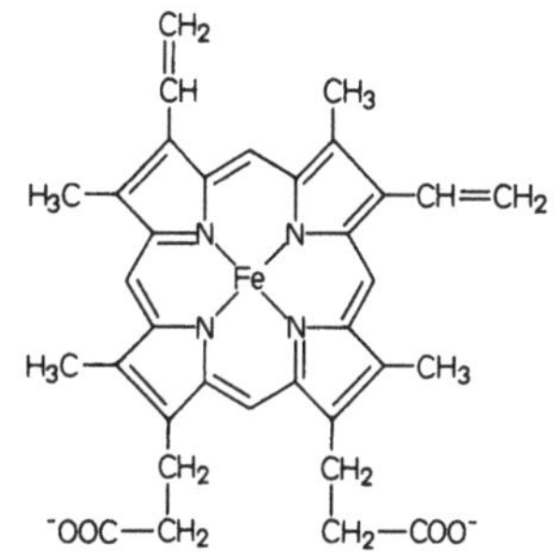

ponents which can be organic or inorganic (or both) in nature. These components are called "prosthetic groups".

The contribution of a prosthetic group to the bulk of a conjugated protein varies greatly. Consider, for example, azurins and myoglobins, both of which have a molecular weight in the neighborhood of 16,000 daltons. The former proteins contain a single copper ion as prosthetic group, the metal contributing about 0.4% of the weight. The myoglobin molecules have iron protoporphyrin IX or "heme" (Fig. 1-1) as prosthetic group, this complicated compound constituting about 4% of the weight. In cytochrome c_3, with three or four heme groups per molecule of weight 12,000 daltons, the prosthetic mass is 17 to 22% of the total.

The terms cofactor, coenzyme, and prosthetic group are not synonymous. If an enzyme requires a nonamino acid derived component for its activity, the component is called a "cofactor". If the cofactor is not simply a metal ion but is a complex organic or organometallic group, it is called a "coenzyme". The "holoenzyme" is the entire, active, enzyme molecule consisting of the cofactor plus the protein moiety. The protein moiety, which by itself is not active, is called an "apoenzyme". The attachment of cofactors to apoenzymes can vary from strong and permanent (covalent) to weak and dissociable. Prosthetic groups are, by definition, stably bound components of conjugated proteins. Thus, some cofactors are prosthetic groups (e.g. the heme group of cytochrome c), some are not (e.g. nicotinamide adenine dinucleotide), and conversely (e.g. hemoglobin is not an enzyme, and the heme prosthetic groups in it are not cofactors).

3. Equilibrium Considerations in Reactions of Transition Metals

a) Site Stability: Equivalent Ligands, Chelation, and Other Factors

In the binding of metal ions to proteins there is competition between these ions and protons for the ionizable residues which serve as ligands. The dissociation of protons from proteins is governed by

$$\log \frac{x_i}{1 - x_i} = \mathrm{pH} - \mathrm{p}K_i + 0.868\ w\ Z$$

where x_i is the extent of proton dissociation, pK_i is the intrinsic pK for group i, and 0.868 w Z is an electrostatic correction factor which depends upon the net charge Z (Tanford, 1961). If the charge on the macromolecule under acid conditions (ionizable groups all protonated) is Z_a, the charge at an arbitrary pH is

$$Z = Z_a + m\bar{v} - R$$

where m is the charge on the metal ion, $\bar{v}$ is the average number of metal ions bound (per molecule), R is the number of protons dissociated, and it is assumed that no other charged species are bound in significant amount. Clearly Z is highly pH dependent. In pH regions where there is neither protein conformational change nor association of protein units, w is approximately constant and the electrostatic correction factor is proportional to the net charge Z. Whether or not this is the case, the following relation describes the association of the metal ion with residue j:

$$\log \frac{\bar{v}_j}{(1-\bar{v}_j)(F-\bar{v})} = \log x_j + \log P + \log K - 0.868\ mwZ$$

where the protein concentration is P, F is the ratio of the total number of metal ions to the total number of protein molecules, and K is the intrinsic association constant between metal ions and the ligand j. The binding of each metal ion eliminates the attached ligands from the equilibrium given in the first equation, a situation explicitly included in the next relation which gives the hydrogen ion dissociation function (titration curve) R:

$$R = \sum_{\text{all } i} \bar{v}_i + \sum_{\text{all } i} (1 - \bar{v}_i) x_i .$$

In the four equations above, the unknowns (each a function of pH) are the x_i, the $\bar{v}_i$, $\bar{v}$, w, Z, and R, of which only R and sometimes $\bar{v}$ can be readily measured. In practice the application of these equations is considerably simplified by the large range of the pK_i (many of the ionizing groups do not have significantly overlapping regions of titration) and by the frequent occurrence of two or more identical ligands in the coordination sphere. If we denote the site consisting of ℓ identical ligands of type j by ℓ_j, the second equation above becomes

$$\log \frac{\bar{v}_{\ell j}}{(1-\bar{v}_{\ell j})(F-\bar{v})} = \ell \log x_j + \log P + \ell \log K - .868\ mwZ$$

since the fraction of these sites free of protons is given by $(1-\bar{v}_{\ell j})(x_j)^{\ell}$.

Consider now the stability of a metal-protein complex against dropping pH under the conditions $F = 1$ (total amount of metal present corresponds to that in the fully formed complex) and $\bar{v}_{\ell j} = \bar{v}$ (ligands involved are only of type j). The preceding equation becomes

$$\log \frac{\bar{v}}{(1-\bar{v})^2} = \ell \log x_j + \log P + \ell \log K - .868\ mwZ$$

which we rewrite in the form

$$(x_j)^{\ell} = \frac{\bar{v}}{(1-\bar{v})^2} \frac{1}{(K')^{\ell} P}$$

where $\ell \log K' = \ell \log K - .868\ mwZ$ defines K'. The value of x_j (the dissociation of free groups j) corresponding to 50% complex formation ($\bar{v} = 1/2$) is then

$$x_j = \frac{1}{K'} \left(\frac{2}{P}\right)^{\frac{1}{\ell}} .$$

The effect of ℓ upon stability can be seen by taking typical values of P (10^{-4} M) and K' (5×10^2). Then for $\ell = 1,2,3,4$ one finds $x_j = 40, 0.28, 0.054, 0.023$, where $(x_j)_{max} = 1$. For these P and K' values the $\ell = 1$ complex cannot ever be fully formed, the $\ell = 2$ complex forms in the pH region where group j titrates in the absence of the metal, and the $\ell = 3$ and 4 complexes are forming at pHs well below the latter titration region. Furthermore, in the region where the change in metal bound with pH is greatest, the buffering capacity dR/dpH is proportional to ℓ^2 (Brill and Venable, 1967). It follows that the slope of the metal association function increases as ℓ^2. In short, as ℓ increases, the apparent pK of the ionizing groups to which the metal ions are bound decreases and the slope of the metal association function increases, the two factors together extending the stability of the complexes to significantly lower pH.

It should be noted that, if a metal ion is on an n-fold symmetry axis, there will be stabilization of the site since it is then necessary for ℓ to be greater than or equal to n.

In the binding of independent small molecule ligands, the entropy decrease accompanying the formation of a link is responsible for a significant positive contribution to the free energy change. When two or more ligand atoms of a molecule are fixed in the proper position to bind to a metal ion, the entropy change is reduced and the association constant increases, a well-known phenomenon called *chelation*. Clearly the quadridentate nitrogen structure of the inner ring of porphyrin is strongly chelating. While, in the absence of a metal ion, the ligands of binding sites on proteins move relative to each other, they are constrained compared with small molecules in solution. Thus binding sites on proteins have at least some chelation character, and stability is gained through this.

Apart from the effects just discussed, the association constant of a complex is influenced by the natures both of the ligands and the metal ion. All the factors involved cannot realistically be combined into a simple mathematical formula, but useful generalizations can be made in such terms as polarizability, "hard" and "soft" acids and bases, and binding order (e.g. the Irving-Williams series for divalent metal ions (Angelici, 1973; Buckingham, 1973)).

b) Electron Transfer and Redox Potentials

The Nernst equation gives the free energy change, expressed as EMF (electromotive force or voltage), for the n electron reduction of an electron acceptor (oxidant):

$$\text{electron acceptor} + ne^- \longleftrightarrow \text{electron donor}$$

$$E = E^{\circ} + \frac{2.303RT}{nF} \log \frac{\text{acceptor activity}}{\text{donor activity}} .$$

$E^{\circ} = E^{\circ\prime}$ is the "standard (redox) potential" corresponding to 1 M concentrations of reductant and oxidant in solution at pH 7.0, 25°C, under 1.0 atm pressure, as measured against a standard (-0.42 volt) hydrogen (gas, 1.0 atm) - hydrogen ion half-cell at the same pH and temperature. At this temperature

$$E(\text{volts}) = E^{\circ\prime} + \frac{0.059}{n} \log \frac{\gamma_a[\text{acceptor}]}{\gamma_d[\text{donor}]}$$

where γ_i, the activity coefficients, are unity under the standard conditions. Closely related to $E^{\circ\prime}$ are the "midpoint" or "half-reduction" potentials, E_m, obtained when the acceptor and donor are at the same concentration. To the extent that the measured EMF and hence the γ_i, are independent of this concentration, $E_m = E^{\circ\prime}$. The more positive these potentials are, the greater the oxidizing potential and, as in an ordinary battery, the greater the affinity of the positive electrode for electrons.

The transfer of electrons is often accompanied by the release or uptake of protons, (as would be associated with a change in the number of groups coordinated to the metal), e.g.

$$\text{electron acceptor} + ne^- \longleftrightarrow \text{electron donor} + mH^+,$$

a process with which is associated a pH dependent EMF of -0.059 m/n volt per pH unit at 25°. Thus potentiometric titration provides a three-dimensional surface on which $E^{\circ\prime}$ (or E_m) is a function of both the concentration of an oxidizing agent and the pH. From such data one can obtain the number n of electrons transferred and the number m of protons involved in the reaction.

In electron transfer reactions involving transition metal valence couples (e.g. cuprous, cupric), n will almost always be 1.0 since the usual case is a unit change in metal valence without any change in ligand oxidation state. No general statement can be made about m and pH dependence. In protein complexes it is not uncommon for m to be different in different regions of pH. Standard (or midpoint) potentials for copper and iron in representative proteins are given in Table 1-2. The couples are either cuprous, cupric or ferrous, ferric so that in each case n = 1.0. Where known, the pH dependence is given. Williams (1976) reviews factors involved in the potentials of one-electron reactions.

Table 1-2. Redox potentials for unit valence changes in metal components of some proteins

Protein	$E^{0'}$ (or E_m) (mV)	$dE^{0'}/dpH$ (mV/pH unit)	
		pH < 7	pH > 7
Azurin (Ps.)	+330 (pH 6.4)		
Plastocyanin	+380	- 60	0
Peroxidases	-190 (average)	see Ricard et al. (1972)	
Myoglobin	+ 50		
Cytochromes b	~ 0 (average)	~ 0	- 60
Cytochromes c	+250 (average)	~ 0	~ 0
Cytochrome a	+210	~20	~20
Cytochrome a_3	+390	0	60
Ferredoxins	-400 (average)		
Rubredoxins	- 60 (average)		
High potential iron-sulfur protein	+350		

References: Dus et al. (1967); Dutton et al. (1970); Hughes (1975); Malkin and Malmström (1970); Palmer (1975); Ricard et al. (1972); Wilson and Dutton (1970); Wilson and Leigh (1972); Wilson et al. (1972a,b)

Transition metals in several proteins (e.g. the cupric ion in azurin and the ferric ion in cytochrome c) are "spontaneously" reduced at alkaline pH (> 10). In the presence of an oxidant such as ferricyanide, cycling between valence states will proceed. Quantitative measurements of the consumption of the oxidant reveal levels of reductant far in excess of the protein concentration (Brill et al., 1968). Not yet known are the identity of the reductant and the nature of the process.

4. Molecular Functions of Proteins Containing Transition Metal Ion Prosthetic Groups

Part of the material discussed in this section can be found in various sections of comprehensive textbooks of biochemistry (e.g. Cantarow and Schepartz, 1967; Lehninger, 1975). Details of topics can be filled in and differences of viewpoint and emphasis explored by consulting more specialized expositions such as those appearing in Antonini and Brunori (1971), Boyer (1975, 1976), Buchanan and Arnon (1970), Burns and Hardy (1975), Chance et al. (1966), Chance et al. (1971), Dunford and Stillman (1976), Florkin and Stotz (1966), Ghiretti (1968), Gray (1971), Hayaishi (1969), King et al. (1973), Lemberg and Barrett (1973), Peisach et al. (1966), Weissbluth (1974). Frieden (1974) discusses relations among the molecular functions of iron and copper proteins and finds that these considerations suggest an evolutionary sequence in respiratory biochemistry.

a) Oxygen Carriage

Substances which are to carry or store molecular oxygen in living systems must have the capacity to take up the gas and to release it under suitable conditions. This function is elegantly met by the well-known protein hemoglobin, a tetramer of molecular weight 65,000 daltons, each subunit of which contains a ferrous protoporphyrin IX prosthetic group. The subunits interact ("cooperate") in such a way that the equilibrium saturation is an S-shaped function of oxygen pressure. Under physiological conditions, this dissociation behavior increases the efficiency of gas loading and unloading by hemoglobin as compared with myoglobin (a monomer with the same prosthetic group) which exhibits the customary hyperbolic dissociation curve. The nature of cooperativity in hemoglobin has been and remains a topic of considerable interest.

The iron-containing proteins chlorocruorin and hemerythrin also transport oxygen. The prosthetic group of chlorocruorin is a heme, but the porphyrin moiety is not the same as in hemoglobin and myoglobin. Hemerythrin is not a heme protein but resembles hemoglobin in having oxy, deoxy, and ferric (met) forms. The hemocyanins are a large class of oxygen-transporting proteins which contain copper (not in a porphyrin complex) and no heme. Vanadium complexes in combination with protein were believed to serve as the carrier of oxygen in sea squirts, but recent data do not support this view.

b) Oxygen Utilization

Depending upon the size and development of the aerobic organism, molecular oxygen will move to the cells requiring it either in solution or else largely carried by one of the proteins just discussed. In small, simple cells (prokaryotes), just within the wall there is a membrane to which oxygen molecules diffuse and where the reduction of these molecules is catalyzed by several kinds of enzyme, with practical consequences (cellular respiration). In complex cells (eukaryotes), the oxygen molecules diffuse inside to the membranous organelles (mitochondria, microbodies) which contain the enzymes of cellular respiration.

Cytochrome c oxidase is found in all aerobic cells. At concentrations of cyanide which effectively inhibit cytochrome oxidase, the oxygen uptake of these cells is severely reduced and death results. This most important enzyme reacts directly with molecular oxygen, transferring to it electrons (received from cytochrome c) and directing the negative free energy made available in this process to the phosphorylation of adenosine diphosphate. Cytochrome oxidase is known to contain two iron atoms,

Fig. 1-2. Heme a. (After Caughey et al., 1975)

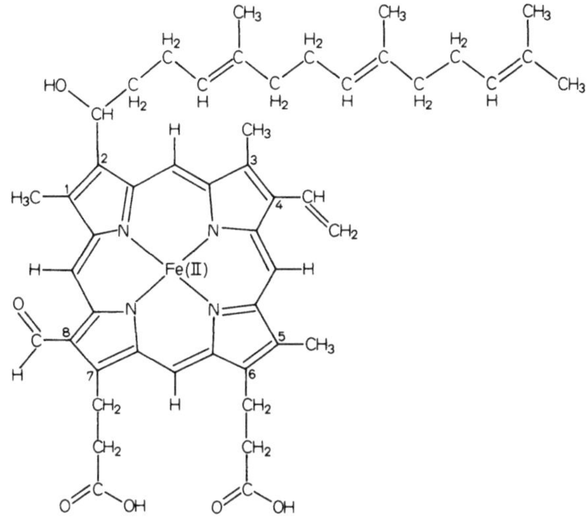

bound in heme a[1] (Fig. 1-2) and two copper atoms. The two hemes are spectroscopically distinguishable in the response of the enzyme to chemical changes and have been designated a and a_3 for some time, where a_3 is the autoxidizable site (Atmungsferment of Warburg). In view of the dependence of the chemical reactivity and spectrally revealed electronic structure of each of these heme sites upon the state of the other, a and a_3 must be regarded as components of one enzyme. Because the reaction with oxygen and succeeding processes are exceedingly fast at room temperature, it has been difficult to resolve functional intermediates in the sequence of electron transfers. However, after flash photolysis of the carbon monoxide-a_3 complex in the presence of dissolved oxygen at -100°C, it is possible to observe recombination of a_3 with the oxygen followed by a succession of intramolecular oxidations (Chance et al., 1975). These experiments indicate that copper is oxidized before "cytochrome a", and hence that the copper sites as well as a_3 and a are components of a single enzyme. Cytochrome oxidase consists of several units, estimates ranging from two to six, and has a total molecular weight of about 200,000 daltons.

The overall reaction of the respiratory chain can be written

$$2NADH + 2H^+ + 6ADP + 6P_i + O_2 \rightarrow 2NAD^+ + 8H_2O + 6ATP$$

where NADH and NAD^+ are the reduced and oxidized forms of nicotinamide adenine dinucleotide respectively (Fig. 1-3). ADP and ATP are adenosine di- and triphosphate, respectively, and P_i is inorganic orthophosphate. Both atoms of oxygen are reduced to water and only electrons are transferred from the NADH. Apart from the (important) occurrence of phosphorylations, all of the enzymes of the respiratory chain could be called electron transferases. Thus cytochrome oxidase engages in electron transport

---

[1]Other heme(s) may be present in certain bacterial cytochrome oxidases.

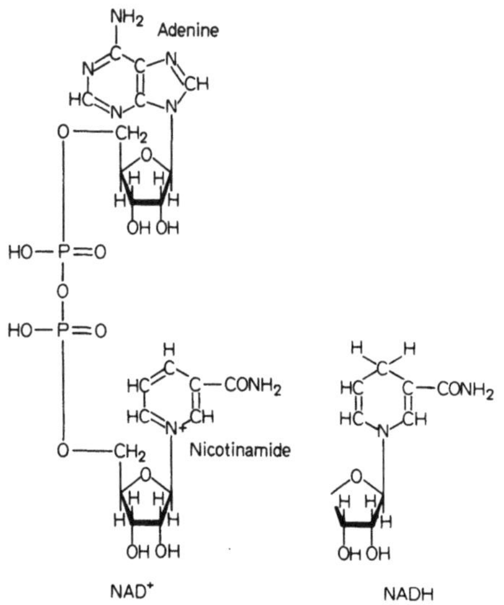

Fig. 1-3. Nicotinamide adenine dinucleotide: oxidized form: *NAD*$^+$; reduced form: *NADH*

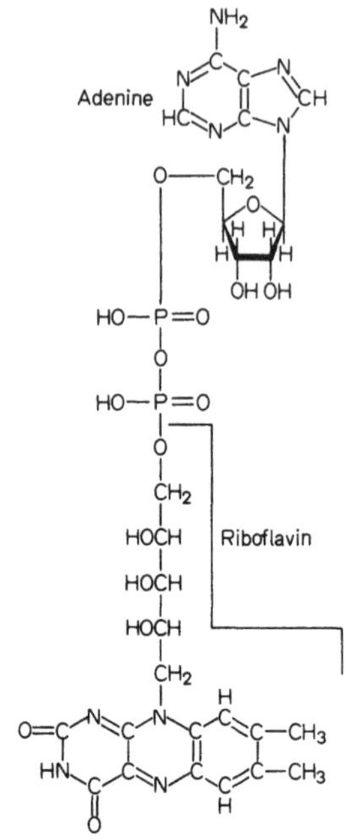

Fig. 1-4. Flavin adenine dinucleotide (FAD)

to oxygen and not in the addition of oxygen to another molecule. There are blue, copper-containing proteins, not in the respiratory chain and not otherwise producing high-energy compounds, that catalyze the transfer of electrons from substrate to molecular oxygen with dehydrogenation of substrate and formation of water. Ascorbic acid oxidase and the laccases are well-established examples. Ceruloplasmin is a blue protein containing seven copper atoms per mole (160,000 daltons), which is found in the blood plasma of most vertebrates. It has recently been established that ceruloplasmin effectively catalyzes the oxidation of the ferrous ion to ferric by molecular oxygen, and the protein is now called ferroxidase. There are a larger number of enzymes which catalyze the transfer of electrons from substrate to oxygen molecules with dehydrogenation of substrate and formation of hydrogen peroxide (or precursor superoxide radicals). This is the case in the reactions catalyzed by the copper-containing enzymes galactose oxidase, uricase, monoamine oxidase and diamine oxidase. The latter two proteins may also contain organic prosthetic factors (flavin or pyridoxal phosphate). The complex protein xanthine oxidase, which contains eight iron atoms, two molybdenum atoms, and two FAD groups (flavin adenine dinucleotide, Fig. 1-4), and has a molecular weight of 300,000 daltons, utilizes molecular oxygen to oxidize certain purines by hydrogen removal, superoxide or peroxide being formed. All of the iron atoms are believed to be present in complexes with an equal number of acid-labile (inorganic) sulfurs. Iron proteins in which the iron is coordinated to sulfur, either inorganic or from cysteine, are now generally called "iron-sulfur" proteins. Aldehyde oxidase is closely related to xanthine oxidase in composition and function. Dihydroorotic dehydrogenase,

Fig. 1-5. Flavin mononucleotide (FMN)

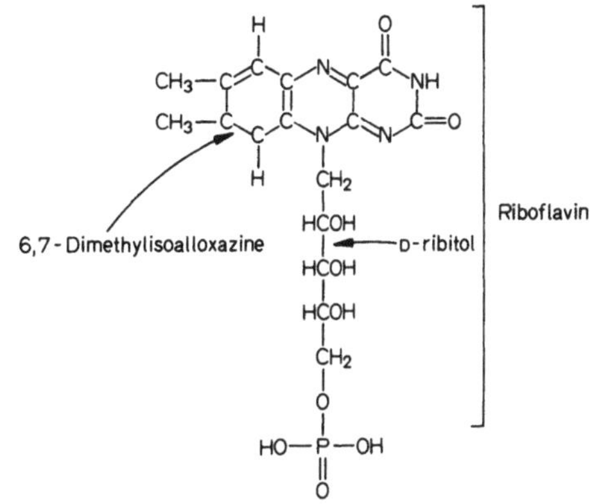

an iron-sulfur protein which contains also both FMN (flavin mononucleotide, Fig. 1-5) and FAD, catalyzes the dehydrogenation of dihydroorotic acid with passage of electrons to oxygen molecules. NAD^+ is a better electron acceptor than oxygen for this system.

In the reactions just discussed, no oxygen atoms are transferred to the substrate. At the other extreme are the oxygen transferases or oxygenases which insert both atoms of the oxygen molecule into the substrate. The iron-containing proteins pyrocatechase, protocatechuic oxidase, homogentisic acid oxidase, and tryptophan oxidase are in this category. In between are the hydroxylases or mixed-function or mono-oxygenases which insert one atom of molecular oxygen into a substrate molecule, the other atom reacting with a separate reducing agent. For example cytochrome P_{450}, found in nonphosphorylating electron transport chains (extramitochondrial) utilizes molecular oxygen to hydroxylate substrates. Other examples are the iron-containing protein proline hydroxylase and the copper-containing proteins tyrosinase and dopamine-β-hydroxylase.

c) Hydroperoxide Reactions

Peroxidases and catalase are heme-containing proteins which catalyze reactions between hydroperoxides (e.g. H_2O_2, hydrogen peroxide; MeOOH, methyl hydroperoxide; EtOOH, ethyl hydroperoxide) and many oxidizable compounds. The defining difference between peroxidases and catalase is in their effectiveness in catalyzing the decomposition of H_2O_2 into oxygen and water, the so-called "catalatic" reaction, catalase being 10^4 more effective. One molecule of catalase cycling through the reaction sequence can free a million molecules of oxygen from hydrogen peroxide (in buffered aqueous solution at room temperature) in one minute. Peroxidases efficiently catalyze the oxidation of hydrogen donors (other than H_2O_2) by H_2O_2. Catalase will do this efficiently only when the level of H_2O_2 is kept relatively low. Catalase from all sources consists of four subunits with a total molecular weight of about 250,000 daltons; four hemes are present. The molecular weights of the peroxidases are variable, averaging about one-fourth that of catalase. The most studied peroxidase is from the horseradish root; it has a molecular weight of 40,000 daltons, of which 18% is carbohydrate, and a single heme group. The heme of catalase and horseradish peroxidase is derived from protoporphyrin IX, but this is not true of all peroxidases.

Catalase and peroxidases react with hydroperoxides to form a green compound (I) which retains both oxidizing equivalents of the peroxide. In the case of peroxidase compound I, these are lost in successive one-relation transfer reactions, with the formation first of a red compound (II) and then the free enzyme. Catalase compound I follows this same path in its reaction with phenols, but with alcohols, formic acid, and nitrous acid, an intermediate compound (II) cannot be detected.

The hydroperoxidase from yeast, which efficiently catalyzes the oxidation of ferrocytochrome c to ferricytochrome c, is in a class by itself. Cytochrome c peroxidase has a molecular weight of 34,000 daltons, contains one ferric protoporphyrin IX and no carbohydrate.

d) Electron Transfer, Protein to Protein

The cytochromes of the respiratory chain exchange iron valence states; some of these reactions are accompanied by phosphorylation of ADP [e.g. cyt b(red) + cyt c(ox) + ADP + P_i → cyt b(ox) + cyt c(red) + ATP + H_2O] and some are not [e.g. cyt c(red) + cyt oxidase (ox) → cyt c(ox) + cyt oxidase (red)]. The pseudo-first order kinetics of the latter reaction at all concentrations of ferrocytochrome c sets it apart from the many enzyme reactions for which Michaelis-Menten analysis is applicable (Smith and Conrad, 1956). While a general rate equation has been established, and mechanisms involving several cytochrome c-oxidase complexes have been found to generate this rate expression, it has not been possible to identify the mechanism which actually is operative (Errede et al., 1976). It is important to note that classification of cytochromes as a, b, c is based upon the positions of absorption maxima in the visible region of the spectrum. The location of the visible bands depends upon the side chains of the particular porphyrin and upon the mode of attachment (iron-protein and porphyrin side chain-protein linkages, and hydrophobic bonds) of the heme to the protein. The latter structural features might be expected to play a role in chemical reactivity, and it is not surprising to find, for example, that cytochrome b is at a lower potential than cytochrome c independent of the organism. The degree to which different cytochromes of the c type share functional abilities is, perhaps, surprising. For example, cytochrome c_2 of purple nonsulfur bacteria under normal conditions can transfer electrons either to a bacteriochlorophyll photoreaction center or to cytochrome oxidase, while eukaryotic cytochrome c is entirely respiratory; yet eukaryotic cytochrome c can donate electrons to bacteriochlorophyll reaction centers in the laboratory. However, on the basis of X-ray analysis and amino acid sequencing of molecules from widely different organisms, many aspects of the structures of all spectrally-designated cytochrome c are the same or very similar (Chap. 2-2) (Dickerson et al., 1976).

The azurins are blue proteins, containing a single copper atom per molecule and no other prosthetic material, which constitute a water soluble fraction of bacterial respiratory cytochrome systems. The azurins exchange valence states with certain cytochromes, no phosphorylation occurring in the process. The organelles containing the photosynthetic apparatus of eukaryotic cells are called chloroplasts. There is, in chloroplasts, an electron-transport chain involved in (photo-) phosphorylation, this chain including cytochromes (related to the b and c classes of mitochondrial cytochromes) and a blue copper-protein, plastocyanin. Chloroplasts also contain as an electron carrier the simplest ferredoxin, a protein of molecular weight about 11,000 daltons (spinach), with two iron atoms bridged by two inorganic sulfur atoms, and each iron externally coordinated to two mercaptides. Ferredoxin is not directly involved in phosphorylation. Prokaryotic photosynthesizing cells are simpler to study than their eukaryotic counterparts because they have only one photosystem. They, too, contain cytochromes, some of which show differences from mitochondrial cytochromes. Photochemical reaction centers in photosynthetic bacteria appear not to have iron-sulfur proteins. Investigations of these reaction centers suggest that an iron-ubiquinone complex is a primary electron acceptor from the bacteriochlorophyll-bacterial pheophytin-ubiquinone chain (Feher et al., 1974), cytochrome being the primary electron donor to bacteriochlorophyll. Nitrogen-fixing and other bacteria utilize complex ferredoxins in electron transport. The electron-transfer protein rubredoxin, also found in microorganisms, contains one iron atom, bound in a tetrahedral mercaptide complex, per molecule of 6100 daltons, or two such complexes in a larger protein.

e) Hydrogen Atom Transfer from Substrate to Coenzyme or Substrate

There is a large class of oxidation-reduction enzymes, the pyridine-linked dehydrogenases, which require either NAD or NADP as coenzyme. In general (glyceraldehyde-3-phosphate dehydrogenase is an exception), these proteins bind and dissociate coenzyme during the catalytic cycle. Hydrogen atoms are transferred from substrate (e.g. lactic acid) to coenzyme in the process. In the case of alcohol dehydrogenase, Zn(II) is known to be a prosthetic group. While many other pyridine-linked dehydrogenases tightly bind divalent metal ions, not all of them do. Transition metal ions are not of significance in this class of enzymes.

The flavin-linked dehydrogenases contain either FAD or FMN, almost always, as prosthetic group. The coenzyme, in general, remains bound to the protein throughout the catalytic cycle. Iron is frequently, but not universally, found to be a prosthetic cofactor in these enzymes. In NADH dehydrogenase (FMN coenzyme) and succinic acid dehydrogenase (FAD coenzyme) molecules there are eight iron atoms, present in the iron-sulfur form. Dihydroorotic dehydrogenase (also mentioned in b above) contains FMN,

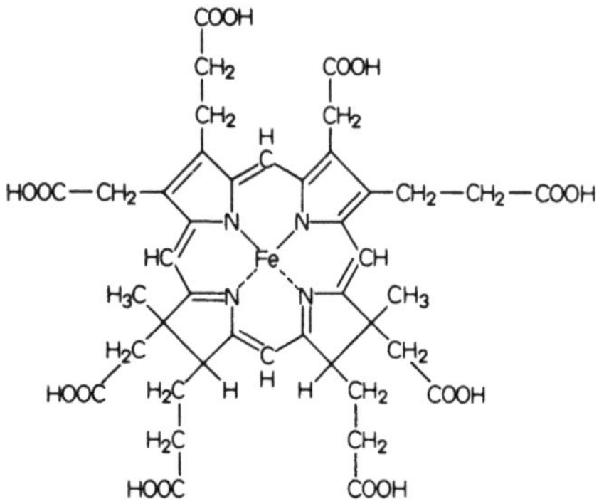

Fig. 1-6. Proposed structure of siroheme. (After Murphy et al., 1973)

FAD, and iron-sulfur complex. The iron in mitochondrial (as distinct from the cytoplasmic enzyme which does not contain iron) glycerol 3-phosphate dehydrogenase, an FAD-linked enzyme, is neither in heme nor the usual iron-sulfur form. Yeast-L-lactate dehydrogenase (cytochrome b_2) has both heme and FMN. *Escherichia coli* NADPH-sulfite reductase and the assimilatory nitrate reductase from *Neurospora crassa* contain siroheme (Fig. 1-6). The latter enzyme requires also both NADP and FAD for activity (Vega et al., 1975).

f) Nitrogen Utilization

The reduction of molecular nitrogen is carried out in nitrogen-fixing bacteria by a complex of two proteins acting in concert. One of the proteins contains iron and molybdenum, and the other just iron. ATP is required for the reaction to proceed. Electrons are supplied by ferredoxin. The iron protein has a molecular weight of about 60,000 daltons, two subunits, and contains four iron and four sulfur atoms, probably in one or two ferredoxin-like structures. The molybdenum-iron enzyme is a four subunit protein, of molecular weight 2,100,000 daltons, with two molybdenum, 24 iron, and 24 sulfur atoms. The iron and sulfur of this protein are believed to be in clusters of the ferredoxin type (Davis et al., 1977).

g) Hydrogen Utilization

Some hydrogenases, bacterial proteins which transfer electrons from molecular hydrogen to an acceptor such as cytochrome or ferredoxin, are known to contain nonheme iron.

h) Hydrogen Peroxide Formation from Superoxide Radicals

The superoxide radical has been found as a product in several oxidase-catalyzed processes. A group of transition metal ion-containing proteins, the superoxide dismutases, catalyze the reaction of superoxide radicals to form hydrogen peroxide and oxygen according to

$$2O_2^- + 2H^+ \rightarrow H_2O_2 + O_2.$$

Many eukaryotic organisms have superoxide dismutase in their cytoplasm. This enzyme consists of two identical subunits of molecular weight 16,000, each subunit containing both one copper and one zinc atom. Superoxide dismutases from mitochondria and from prokaryotic cells have been found to contain manganese or iron in molecules consisting of two or more subunits each of weight about 20,000 daltons. The copper and zinc superoxide dismutases have similar physical and chemical properties (e.g. activity is cyanide-sensitive), and the manganese and iron superoxide dismutases also have in common certain structural and chemical properties (e.g. activity is cyanide-insensitive) (Fridovitch, 1975).

Table 1-3. Occurrence of certain transition metal ions in proteins which perform specific molecular functions

Molecular function	Fe, not in a porphyrin complex	Fe, in a porphyrin complex	Cu	Mo	Mn
Oxygen carriage	+	+	+	-	-
Oxygen utilization	+	+	+	+	-
Hydroperoxide catalysis	-	+	-	-	-
Electron-transfer protein to protein	+	+	+	-	-
Hydrogen atom transfer	+	+	-	-	-
Nitrogen utilization	+	-	-	+	-
Hydrogen utilization	+	-	-	-	-
Superoxide radical catalysis	+	-	+	-	+

The occurrences of iron, copper, molybdenum, and manganese in proteins which carry out the molecular functions discussed above are summarized in Table 1-3. The entry + signifies that the metal ion is present in at least one protein with the specified function, and - indicates that there is no protein with the specified function which is (at the time of compilation) known to contain the metal ion. The extents of occurrence can very greatly among the active members of a row; for example, with regard to oxygen transport, iron in hemes is very widespread, copper is well represented, but iron not in a porphyrin complex is found only in hemerythrin.

i) Methyl Transfer and Internal Hydrogen Transfer

The cobalt-containing vitamin B_{12} enzymes catalyze specific group transfers. The reader is referred to Wood and Brown (1972) for a review of the vitamin B_{12} enzymes.

j) Utilization, Production, and Transfer of Ammonia

Evidence for the involvement of transition metal ions, especially nickel, in the mechanisms of action of enzymes using ammonia or amide is reviewed by Dixon et al. (1976).

k) Oxygen Evolution

Manganese is essential to the normal functioning of photosystem 2 and necessary for the evolution of oxygen in plants (Cheniae and Martin, 1971, 1972). The manganese in the oxygen-evolving system appears to be loosely bound since most of it can be extracted by washing with hydroxylamine. Association of the metal with a protein has not been established but is very likely in view of the requirement for enzyme function.

5. The Role Which a Transition Metal Ion Plays in the Function of a Protein

Let us assume that a protein has been purified and that chemical and physical analyses have given the number of each kind of transition metal ion binding site. Taken together, answers to the following questions will describe the role which the transition metal ion plays in the mechanism of action of the protein.

1. What is the electronic configuration of the metal ion, and what is the spin state (Chap. 6-2) (if more than one spin state is compatible with the configuration)? (Knowledge of the electronic configuration implies knowledge of the valence state, but the converse is not necessarily true.)

2. To how many atoms is the metal ion bonded, and how are they disposed (symmetry)? (The groups contributing the bound atoms are called ligands.)

3. What is the chemical nature of each bound atom? (By chemical nature one means, for example, an imidazole nitrogen of a histidine residue.)

4. How covalent are the bonds? (How delocalized are the metal and ligand electrons?)

5. What are the bond lengths?

6. How do the answers to (1) through (5) change as the protein carries out its function?

6. Experimental Methods

One attempts to answer the above questions with the help of many experimental techniques, several of which are discussed below.

a) Isolation and Identification of Prosthetic (Nonprotein) Ligands

The heme group can be separated by chemical means from the protein moiety of hemeproteins and readily (except for a few porphyrins with unusual side chains) analyzed to yield information relative to (2), (3), and (5) above (Falk, 1964).

b) X-ray Diffraction from Protein Single Crystals

This is the only method which can give the three-dimensional structure of the entire protein molecule at atomic resolution. In conjunction with additional knowledge of the sequence of amino acid residues along the polypeptide chain(s) obtained by identification of short peptides from the hydrolyzed protein, the X-ray diffraction-derived electron density map enables one to locate the positions of essentially all the atoms (with the exception of protons) of the protein - including, of course, those of the prosthetic group(s). In this manner, complete answers to questions (2) and (3) are obtained. Question (5) is answered too, but with a resolution that may not be sufficient to detect differences in structure that could be important in biochemical reaction mechanisms. In terms of the quality of the X-ray diffraction data from protein crystals, there is a lower limit of about 1 Å to the resolution of structure in the electron density map. It is possible to perform a least squares fit of segments of "known" structure (e.g. aromatic side chain, stretch of α-helix, porphyrin group, dimensionally specified in whole or in part) to the electron density map in order to determine other coordinates with greater precision than 1 Å, but the result contains whatever uncertainties there are in the assumed substructure(s). With X-ray diffraction, the structure of small molecules is resolvable to about 0.01 Å. Thus, for example, it is well known that carbon-carbon single bonds are all close to 1.54 Å in length, carbon-carbon double bonds, 1.35 Å, and the carbon-carbon bond in benzene-like aromatic rings, 1.40 Å. Differences in length of less than 0.1 Å are significant when a description of the electronic structure of a bond is sought.

The problem of obtaining X-ray diffraction data from reaction intermediates (e.g. enzyme-substrate complexes) remains to be solved in a general way. However, some enzyme-inhibitor compounds, and even slowly reacting (cross-linked) enzyme-substrate compounds, have been studied (Steitz et al., 1967; Wyckoff et al., 1967).

c) Electron Paramagnetic Resonance and Related Measurements

Electron Paramagnetic Resonance Spectroscopy (EPR). EPR is an especially direct probe of structure in the immediate neighborhood of a paramagnetic transition metal ion. As with X-ray diffraction, orientation studies provide the maximum amount of information: (1) the number of distinct magnetic sites; (2) the orientation with respect to the crystal axes of symmetry axes of each magnetic site; (3) the symmetries of the environments of the magnetic sites; (4) the ground states of the transition metal ions; (5) the delocalization of metal electrons onto the ligands; and (6) identification of the ligand atoms bound to the metal ions when the former have nuclear magnetic moments (Brill et al., 1971). With solutions and polycrystalline samples, clearly (2) cannot be measured, and other information can be obscured. In general, however, whole or partial answers to questions (1), (2), (3) and (4) can be obtained with EPR. It is also possible to employ EPR linewidth studies to provide bond lengths and additional information about the chemical nature of ligand groups (Brill and Venable, 1972).

It is often necessary (e.g. with hemeprotein compounds) to conduct EPR measurements at cryogenic temperatures in order to obtain a signal or to improve the resolution of a spectrum. In the case of heme, this departure from physiological conditions is sometimes accompanied by non-Curie law dependence of the magnetic susceptibility upon the temperature, and by temperature-dependent spectroscopic changes (Iizuka and Kotani, 1968, 1969; Tamura, 1971a,b; Tasaki et al., 1967). This temperature behavior can usually be explained in terms of thermal equilibria between high-spin and low-spin states of the ferric ion, and, at very low temperatures, in terms of the zero-field splitting of the high-spin state (Chap. 6-6). However, there are effects observable near the freezing point which do not have a simple explanation (Ehrenberg and Estabrook, 1966; Iizuka and Kotani, 1969). While many proteins are found to be not harmed by freezing, others are; and protein crystals suffer varying degrees of disorder upon freezing and thawing. For any measurement made with the sample at low temperatures, the system under study should be examined for possible structural changes in the protein and chemical changes in the solvent which can occur upon freezing and/or attend further lowering of the temperature.

Rapid freezing is a useful tool for stopping a reaction at specified times after initiation, the measurements then being made upon the frozen samples at leisure. This technique has been developed by Bray (1961) and Bray and Pettersson (1961), and employed to study enzyme-catalyzed reactions by EPR, thereby obtaining answers to question (6), subject to the precautions outlined in the preceding paragraph. EPR spectroscopy can also be carried out on reacting solutions at room temperature (Yamazaki et al., 1960).

Electron-Nuclear Double Resonance (ENDOR). In the ENDOR measurement a radiofrequency region is swept while a saturated EPR signal is monitored (at fixed microwave frequency and magnetic field strength). Excursions of the EPR amplitude can be observed when the radiofrequency passes through Bohr frequencies corresponding to transitions between hyperfine levels of paramagnetic ligand nuclei coupled to the electron spin. ENDOR provides answers to questions (3) and (4), and its application to transition metal ion-containing proteins promises to be fruitful (Rist et al., 1970; Scholes et al., 1972; Scholes and Van Camp, 1976).

Magnetic Susceptibility. The primary usefulness of the magnetic susceptibility measurement is in the determination of the spin state of the transition metal ion. This information is essential in arriving at an answer to question (1), or in confirming an answer obtained by other means. It is also often necessary to know the spin state in order to be able to interpret EPR spectra. From the temperature dependence of the magnetic susceptibility at low temperature, it is possible to evaluate zero-field splittings. For two-component systems, each component of known magnetic moment, the susceptibility measurement will give the composition, provided there is no significant interaction between the components. For systems of known composition, the magnetic susceptibility can reveal the existence of antiferromagnetic exchange coupling when other methods have provided the spin states of the component atoms. The susceptibility measurement can be made on a rapidly reacting system (Brill, 1969).

d) Optical Measurements

Light Absorption. There is a body of knowledge, largely empirical but supported by theory, which qualitatively relates the positions and intensities of light absorption bands to the valence and spin state of a transition metal ion and the geometrical arrangement and chemical nature of its ligands. Discussions of this kind of information for copper and iron can be found, for example, in the reviews of Braterman et al. (1964), Brill et al. (1964) and Brill (1966). In most cases, light absorption spectroscopy by itself cannot be said to provide quantitative or detailed answers to questions (1), (2), and (3), but indicates rather the range of possible answers. Given a model for a transition metal ion complex, one can calculate the area under an absorption band (Chap. 6-5) and compare this "oscillator strength" with the experimental value. Difference spectroscopy in the region of amino acid residue absorption (deep ultraviolet) can be employed to help in the identification of protein-contributed ligands (Brill and Sandberg, 1968; Sandberg and Balegh, 1973). The light absorption measurement can be made during fast biochemical reactions, yielding information related to question (6) with exceptionally fine time resolution (Roughton and Chance, 1963).

Optical Activity (Circular Dichroism). The optical activity of a metal ion chromophore depends in a sensitive way upon the stereochemistry of the binding site. The measurement of optical rotatory dispersion (ORD) has been supplanted by that of circular dichroism (CD), because the overlap of bands in the CD spectrum is significantly reduced compared with the ORD spectrum (thereby, for example, essentially eliminating the background contributions to the visible region from protein chromophores). The area under a CD band can be compared with the "rotational strength" calculated from a model to test the validity of the model. The rotational strength depends upon the magnetic as well as the electric dipole character of a transition. While absorption bands are always positive, CD bands can be either positive or negative. For this reason, CD is often helpful in resolving bands which because of overlap cannot be distinguished in the absorption spectrum.

The combination of light absorption and CD measurements can provide significantly more information of the kind sought in questions (1), (2), and (3) than either measurement alone. These in turn are complemented by the EPR spectrum and any model of the binding site must be consistent with all three kinds of data. Indeed the association of particular structural features solely with a particular measurement breaks down at some level of quantitative treatment. All of the parameters which provide numerical answers to questions (1) through (5) enter calculations of both electric and magnetic dipole matrix elements. Because these matrix elements appear in different ways in the expressions which describe the EPR, absorption, and CD bands, the three measurements provide independent checks of the (quantitative) model, giving the various structural features different weights (Blumberg, 1966; Brill and Bryce, 1968).

The protein moiety exhibits optical activity in the ultraviolet which has been well studied and described (Crabbé, 1972; Jirgensons, 1973). Planar porphyrins are not themselves optically active, but can acquire optical activity when asymmetrically perturbed and when porphyrin and metal or protein (e.g. aromatic side-chain) transitions are coupled together (Hsu and Woody, 1971). Since the optical activity induced in heme depends largely upon the protein structure in the immediate environment of the prosthetic group, the CD (or ORD) can reflect changes in metal coordination which have altered the local protein conformation [questions (2) and (5)].

Magnetic Circular Dichroism. Whether a substance is naturally optically active, optical activity can be induced by the application of a magnetic field, an effect discovered by Faraday. The measurement of magnetic circular dichroism (MCD) provides the same information as magnetic optical rotatory dispersion and has several advantages (Schatz and McCaffery, 1969). The MCD associated with an optical transition depends upon specific details of the ground and excited states, e.g. degeneracies, and is insensitive to changes in the protein which do not directly affect the structure (energy levels) of the chromophore. Wider application of MCD in metalloprotein research is suggested by the results which are at hand (Coleman and Coleman, 1972; Cheng et al., 1973; Sutherland and Klein, 1972; Thomson et al., 1976; Treu and Hopfield, 1975; Vickery et al., 1976a,b) [questions (1) and (2)].

Resonance Raman Spectroscopy. If a monochromatic beam falls upon a collection of molecules, the scattered energy will mostly be of the same wavelength as the incident radiation, but "Raman" bands will also appear at wavelengths above and below. The Raman components arise from modulation of the molecular electric polarizability by molecular vibrations. Resonance Raman spectroscopy refers to measurement of the spectrum of scattered light when the excitation is within an absorption band of the chromophore being studied. With a laser light source, resonance Raman spectroscopy provides a sensitive means of obtaining information about (unresolved) vibronic components of electronic absorption bands (Spiro, 1975; Spiro and Strekas, 1972) and about the atoms involved in ligand-metal bonds (Miskowski et al., 1975), providing information related to questions (2) and (3).

X-ray Spectroscopy. Synchrotrons provide a source of X-rays which is continuous in energy and intense. This synchrotron radiation can be used to measure the X-ray absorption spectra of transition metal ions in compounds, including proteins. At the K edge (transitions from the 1s state to bound states) these spectra show differences in atomic transition energies among compounds. For a given absorbing atom, such differences arise mainly from variations in the charge on the atom, and reflect the oxidation state of the atom and the distribution of bonding electrons [questions (1), (3) and (4)]. The intensities of the "forbidden" transitions provide information about the site symmetry [question (2)]. Useful applications of K edge spectroscopy to iron group complexes, including iron-sulfur and heme proteins, have been reported (Shulman et al., 1976). Absorption from the adjoining higher energy regions ("extended X-ray absorption fine structure" or EXAFS) has the potential for providing interatomic distances [question (5)] with a fine resolution (Kincaid et al., 1977; Stern, 1974).

X-ray photoelectron spectroscopy (XPS) can be employed in identifying ligands [question (3)] (Solomon et al., 1975).

e) Measurements Utilizing Nuclei

Nuclear Magnetic Resonance (NMR). The improvement in resolution attending the use of superconducting solenoids to produce the polarizing field for NMR spectroscopy has made this measurement effective in the study of biological macromolecules (Phillips, 1971; Wüthrich, 1970). Thus far most of these studies have dealt with proton resonances, but improvements in experimental technique (Fourier transform spectroscopy) and data processing (computer averaging) are increasing the number of nuclei which can be used as probes. The NMR measurement is usually done on liquid solutions, in which case the parameters so obtained are spatial averages (isotropic components). When a paramagnetic metal ion is part of the molecule being examined by NMR, resonances can be shifted relative to their position in the analogous diamagnetic molecule. Phillips (1973) has reviewed the occurrence and analysis of these "contact shifts" in proton magnetic resonance (PMR) spectra from biological molecules. The contact shifts are of two kinds, hyperfine (Fermi) (Chap. 6-2 and 6-7) and pseudocontact. The former arises from, and is proportional to, the unpaired electron spin density at the nucleus exhibiting the effect [question (4)]. The latter is dipolar in origin, requires for its existence anisotropy in the magnetic susceptibility

(Chap. 4-2) of the paramagnetic metal ion, and falls off as the third power of the distance from the ion [question (5)]. Pseudocontact and hyperfine contact shifts exhibit the same temperature dependence (usually Curie law, but otherwise for antiferromagnetically coupled metal ion systems), and the contributions from the two can be difficult to separate experimentally. Realistic calculation of the pseudocontact shift requires considerable experimental information (magnitudes and directions of the g-values of the paramagnetic metal ion, its spin state, and the coordinates of the nucleus) (Horrocks and Greenberg, 1973; Shulman et al., 1971). Apart from complex analytical procedures, if a change in metal ion spin produces a major change in a proton resonance assigned to a potential ligand (e.g. histidine), this suggests direct coordination of the ligand to the metal [question (3)] (Markley et al., 1975).

Nuclear Spin Relaxation Rate. Water and ligand exchange onto protein-bound paramagnetic metal ions can be studied by nuclear spin relaxation rate measurements. These measurements yield, with varying degrees of completeness and accuracy, information of types (2), (3) and (5) (Mildvan and Cohn, 1970). Thus far, this useful method has been more broadly applied to systems with paramagnetic probes extrinsic to the protein than to proteins which naturally contain metal ions, although the results from the latter studies have been helpful. For example, relaxivity studies have shown that the iron atoms in cytochrome *c* and in iron-sulfur proteins are not accessible to water protons, implying that the metal is shielded from direct interaction with substrates and hence that electron transfer reactions are not of the inner sphere type (Chap. 1-7).

Lanthanide Shift Reagents. Extrinsic lanthanide probes can produce large pseudocontact shifts of proton resonances in organic molecules and/or broaden the resonances. Chemically very similar lanthanides have useful differences in electronic relaxation time (hence differences in the relative effectiveness in producing contact shift as compared with line broadening). Taken together, the changes in the proton magnetic resonance spectrum produced by two or more lanthanides bound at the same site can be used to obtain estimates of intramolecular distances (shifts go as r^{-3}, widths as r^{-6}) and angles. Lanthanide probes are being used effectively to study the structure of proteins in solution (Campbell et al., 1975).

Mössbauer Spectroscopy. Mössbauer spectroscopy is restricted to the ^{57}Fe nuclide among isotopes of metal atoms of biological interest. The spectrum of gamma ray absorption by ^{57}Fe in the molecule of interest is measured. The first excited state of the nucleus has a spin of 3/2; the resulting quadrupole splitting (Chap. 6-2) depends upon the electric field gradient at the nucleus and hence reflects changes in the valence electron distribution. The mean square radius of the first excited state differs from that of the ground state; the transition energy then is proportional to the (s) electron probability density at the nucleus which, once again, depends upon the distribution of the electrons involved in bonding. The Mössbauer measurement provides information related to questions (1), (2) and (4) (Bearden and Dunham, 1970; Lang, 1970; Moss, 1973).

7. Some Aspects of the Role of the Polypeptide in the Functioning of Proteins Containing Transition Metal Ions

The binding of metal ions directly (without nonamino acid intermediate groups) to protein often occurs, e.g. in the blue copper proteins. Apart from the possibility of bound water, the coordination "sphere" is then a protein "surface" or "pocket". Just as there are relatively few kinds of prosthetic group, the number of functional, wholly protein, metal-binding sites which differ fundamentally (not just by conservative, chemically similar substitutions) in structure may be much more limited than might be supposed from the many possible sites suggested by model building. That is, there may be broad representation (although unlikely to be as widespread as that of porphyrin prosthetic groups) of homologous polypeptide loops of ten to twenty residues which serve to bind metal ions in some unique manner (Kretsinger, 1975). An extension of this idea applies to the binding of more complex prosthetic groups. X-ray structures and amino acid sequences are available for members of the spectrally designated c type cytochromes from eukaryotic sources and from photosynthetic and respiratory bacteria. In all these c cytochromes, the immediate environment (coordination) of the iron is alike, as is the covalent attachment of the porphyrin to the polypeptide chain (Chap. 2-2). Furthermore, where available, high resolution structures show that the polypeptide chain is folded around the heme so as to leave open to the solvent phase just one edge, with positively charged lysines at the ends of the opening in the protein. While there are differences at the surface of these molecules (which account for differences in the chain lengths - 103, 112, and 134 residues for tuna c, *Rhodospirillum rubram* c_2, and *Paracoccus denitrificans* c_{550}, respectively), the interiors, including the hydrogen and hydrophobic bonding of the porphyrin side chains to the protein, are the same. It has been proposed that all of the c (and b) type cytochromes belong to electron transport chains (now functioning in various photosynthetic and respiratory pathways) that are evolutionarily homologous (Dickerson et al., 1976).

Interactions extending over the greater part of the macromolecule can play a role in the function of a protein through effects of the three-dimensional conformation upon the local environment of a special site. This view is crystallized in the hypothesis of an "entatic" nature of active sites: the catalytic ability of enzymes derives not from unusual chemical groups, but rather is conferred upon groups by the surrounding structure which gives the ground configuration characteristics of a transition (activated) state (Moore and Williams, 1976; Vallee and Williams, 1968a,b). Considerable support for an entatic component in the action of catalytic proteins comes from metalloenzymes wherein the spectroscopic and magnetic properties of the ground state of the metal ion are often unusual (see, e.g. Chapter 3-3). The protein, however, cannot be pictured as statically fixing the environment at the active site. Many of the interactions which determine the average conformation of a protein are not much larger than thermal energies, and, even in the absence of reactions with substrates and external ligands, the structure of proteins in solution is dynamic. Temporal fluctuations in protein structure have been studied by hydrogen exchange kinetics and quantitatively described in terms of the free energies of "partially denatured forms" in equilibrium with "the native form" (Englander, 1975). Thus at least some groups in these macromolecules continuously pass through a range of states; for the active sites of enzymes there is the implication that among these states are highly reactive ones. In this connection it is of interest to note that the heme plane can be readily ruffled, domed, or both (Hoard, 1971).

Turning to roles other than prosthetic group binding, one finds strong evidence for the involvement of specific residues, not linked to heme iron, in the mechanisms of action of catalase and peroxidase. Aminotriazole reacts with compound I of catalase, adding to an imidazole group and irreversibly inhibiting the enzyme (Agrawal et al., 1970; Margoliash and Schejter, 1962). Since only small spectral changes occur upon formation of the aminotriazole derivative from free catalase, the histidine residue attacked is unlikely to have been bound to heme iron. Kinetic studies of the inactivation of horseradish peroxidase by high concentrations of hydrogen peroxide (Weinryb, 1966) and by azide (Brill and Weinryb, 1967) demonstrate the participation of a nonheme site as well as a heme site in the catalytic process, and the azide results suggest the involvement of a methionine residue in the mechanism of action.

Early evidence for the transfer of an electron from an internal group to the protein outer surface is found in the kinetic measurement of Chance (1951) on the reaction of cytochrome c with cytochrome c peroxidase which revealed a rate in excess of the collision frequency between the iron atoms. The implication that electron transfer can occur at the periphery of a protein molecule is supported by later structural, immunological, kinetic, and spectroscopic studies (Chaps. 2-2, 2-4, 3-4). Electron tunneling in other processes is inferred from kinetic measurements at cryogenic temperatures (Chap. 4-10). Free particle states of electrons are unlikely to be involved in these processes. Rather the electron is transferred from a bound state of one molecule to that of another. Hopfield (1974) has presented an intermolecular, short range (~ 10 Å), tunneling model in which the vibrational energies of the individual molecules determine the temperature dependence of the rate of electron transfer. In the latter scheme there is a characteristic temperature $T_c \approx \hbar\omega/k$ ($\hbar\omega$ is a vibrational quantum, k is the Boltzmann constant) such that below $T_c/2$ the tunneling rate becomes largely temperature-independent. A role for the protein in the temperature dependence of tunneling has been suggested which allows, through thermal contraction, either an increase or decrease in electron transport when the temperature is lowered, as has been found experimentally (Hales, 1976; Hopfield, 1976).

Electron-transfer reactions of small metal complexes in solution are often described in terms of inner-sphere and outer-sphere processes (Taube, 1970). In an inner-(really coordination-) sphere reaction, an intermediate is formed in which two metal ions are connected to a bridging ligand. Outer-sphere electron transfer does not involve a common bridging group, the donor and acceptor metals reacting through a chain of interposed atoms; the donor and acceptor orbitals overlap weakly, and this process is closely related to tunneling (Jortner, 1976). Studies of the reduction of ferricytochrome c by relatively small inorganic complexes suggest that outer sphere reactions dominate (Hodges et al., 1974; Wherland and Gray, 1976). This may not be the case for the oxidation of ferrocytochrome c by ferricyanide (Ilan et al., 1976). The reduction and oxidation of cytochrome c by small inorganic reagents appear, in these studies, to take place at a common site involving a cluster of amino acids near a heme edge. This is consistent with the molecular structure of cytochrome c, in which the metal ion has severely limited access to the aqueous phase but one edge of the porphyrin ring is exposed.

Chapter 2

Metal Coordination in Proteins

1. Ligands

Water and buffer ions are always available and frequently involved as ligands. In particular, the oxygen atom of the water molecule forms a stable coordinate link with iron group ions. Putting aside these very small and unconstrained ligands, we turn to the metal-binding groups on proteins and associated with proteins.

Among the ionizable amino acid residues, histidine, through nitrogens of the imidazole ring, is of central importance in metal coordination (Sundberg and Martin, 1974). In proteins for which the environment of the metal atoms has been established by X-ray structure analysis and chemical sequencing (Chap. 2-2), there are many cases where one or more histidine residues serve as ligands. The nitrogen involved loses a proton as the pH is raised toward neutrality (Fig. 2-1) and is then free to bind a metal ion. Amino nitrogens also coordinate strongly with metal ions, although these nitrogens are somewhat different electronically than imidazole nitrogens (Chap. 6-4). At least one terminal α-amino group is always present and provides a nitrogen which dissociates a proton at somewhat alkaline pH. Lysine is usually present and has an ε-amino group, the nitrogen of which does not become available for metal binding until still higher pH.

Carboxyl groups, when ionized, have negatively charged oxygens available for coordination. These ligands are provided at C-terminals and by the dicarboxylic acid residues aspartic and glutamic. While it ionizes at a very high pH, tyrosine can provide phenolic oxygen for metal binding. Similar with regard to pK is cysteine, which dissociates a sulfhydryl proton, making a negatively-charged sulfur available for coordination.

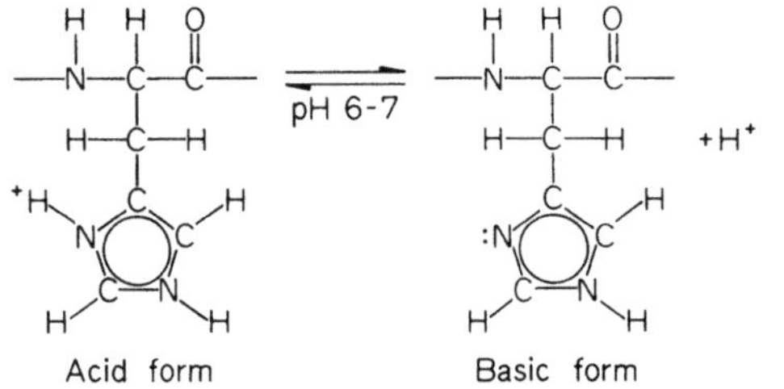

Fig. 2-1. Dissociation of proton from imidazole ring of histidine residue at neutral pH

Table 2-1. Ionizable protein groups which participate in metal binding

Ionizable group	Location or residue	Intrinsic pK
α-COOH	Carboxyl terminal	3.5 -4
β,γ-COOH	Aspartic, glutamic	4 - 5
Imidazole	Histidine	6 - 7
α-NH_3^+	Amino terminal	7.5 - 8
Sulfhydryl	Cysteine	10
ε-NH_3^+	Lysine	10
Phenolic	Tyrosine	9.5 - 10.5
Guanidine	Arginine	12

Table 2-2. pH midpoints for deprotonation from glycylglycine in presence of metal ions.(After Martin, 1974)

Metal ion	^+H_3N-CH_2-CO-NH-CH_2-COO^- ↓	↓
None, pK_a	8.1	~13
Co(II)	7	10
Ni(II)	6.5	9.7
Cu(II)	4.2	4.8

The ionizable protein groups which participate in metal binding and the corresponding pK (intrinsic) for the appropriate ionization are listed in Table 2-1. The pH at which deprotonation is 50% complete depends upon the charges on all the ionizable groups of the protein, normally being on the acid side of the intrinsic pK when the solution is acid to the isoelectric point of the protein, and conversely. The reader is referred to texts on the physical chemistry of macromolecules and/or polyelectrolytes (e.g. Tanford, 1961) for quantitative treatment of these electrostatic effects.

For all ionizing ligands, the pH at which metal complex formation (involving the proton-free group) proceeds can be several log units lower than the pK for hydrogen ion dissociation (in the absence of the metal). This promotion of ionization and complexation depends upon the stability of the entire set of bonds in the structure so formed. The pHs at which protons and metal ions compete with equal success for binding to the nitrogens of glycylglycine are given in Table 2-2. Note that the amide group will dissociate a proton in favor of linking to a metal ion. A sizeable portion of the literature on peptide-metal complexes deals with this coordination situation, one which is unlikely to occur within the physiological range of pH except in the case of the cupric ion.

Among nonionizable residues, methionine is linked, via the sulfur atom, to heme iron in cytochrome c. Serine has not been found bound to a metal ion.

In the hemeproteins, which constitute a sizeable fraction of all iron-containing proteins, four of the coordinate positions of each metal ion are occupied by the pyrrole nitrogens of a porphyrin, most often protoporphyrin IX (Fig. 1-1). In free base porphyrins, two of these four nitrogens are protonated. Formation of a metalloporphyrin is accompanied by the loss of the latter protons so that, as far as the coordination sphere is concerned, the quadridentate porphyrin ligand bears a negative charge of two units. Protoporphyrin IX also contains ionizable side chains, two propionic acid groups which add electrostatic bonding to the already strong hydrophobic interactions which retain the porphyrin in its place in the protein molecule. This synergistic picture is completed with the inclusion of an imidazole-iron link in most hemeproteins (cytochrome b_5 has two such links and cytochrome c has a methionine sulfur-iron as well as an imidazole-iron link).

2. The Established Coordination in Several Proteins

Various features of metal ion coordination in several proteins as established by chemical, X-ray diffraction, and spectroscopic methods are summarized in Table 2-3. From this table one sees that in each case nitrogenous bases constitute not less than one-half of the ligands. Imidazole nitrogens are more effective than pyrrole nitrogens in securing a metal ion to the protein moiety since the imidazole ring is covalently bonded to a polypeptide chain and porphyrin usually is not. (In cytochrome c two of the side-chains of the heme group are also covalently bonded to cysteine residues of the protein.) Metal coordination in the iron-sulfur proteins, described in h and i below, is very different than in any of the proteins of Table 2-3.

a) Insulin (Adams et al., 1969; Blundell et al., 1971; Brill and Venable, 1968, 1972)

Insulin monomers (Sanger units) of molecular weight 6000 daltons interact with each other to form very stable dimers. Three of these dimers will bind a pair of zinc (or other divalent 3d) ions. This association is strong and orderly (crystals readily form). One metal ion serves as a common linkage point for the B10 histidine residues of three monomers (each belonging to a different dimer), and the other metal ion, for B10 residues of the other three monomers. Here again is synergistic structural stabilization. By virtue of independent protein subunits contributing ligands, there is little constraint imposed upon the final geometry of the complex, and the coordination symmetry is high, as it is with small complexing ions or molecules. Thus, the coordination of zinc in insulin crystals is found to be distorted octahedral (there are three water molecules as well as the

Table 2-3. Features of metal ion coordination as established in several proteins

Protein	Metal ions	Porphyrin	Protein-contributed ligands	Waters	Coordination number	Total nitrogens	Total oxygens	Delocalization $(\alpha')^2$ sigma bond
Insulin 2M(II)	Zn(II) or Cu(II)	no	3 histidines	3	6	3	3	0.31 (from Cu(II))
Insulin 4M(II)	1 Zn(II)	no	3 histidines	3(?)	6	3	3(?)	-
	3 Zn(II)	no	2 histidines	2	4	2	2	-
Carboxypeptidase[a]	Zn(II) or Cu(II)	no	2 histidines 1 glutamate	1	4	2	2	0.28 (from Cu(II))
Myoglobin, Hemoglobin	Fe(II) or Fe(III)	yes	1 histidine	0 or 1	5 or 6	5	0 or 1	0.31 (from Fe(III)Mb)
Myohemerythrin, Hemerythrin	Fe(II) or Fe(III)	no	2 histidines, 1 tyrosine, 1 aspartate, 1 glutamine; 3 histidines AB: 2 histidines, 1 tyrosine, 1 aspartate BC: 1 aspartate, 1 glutamine, 3 histidines	1	AB: 6 BC: 6	AB: 2 BC: 3	AB: 4 BC: 3	-
Cytochrome c	Fe(II) or Fe(III)	yes	1 histidine 1 methionine	0	6	5	0(S)	-
Cytochrome b_5	Fe(II) or Fe(III)	yes	2 histidines	0	6	6	0	-
Cu, Zn superoxide dismutase	Cu(II) and	no	4 histidines	0	4	4	0	-
	Zn(II)	no	3 histidines 1 aspartate	0	4	3	1	-

[a]The cupric form of carboxypeptidase is inactive. It is listed here because it provides a value for sigma bond delocalization

three imidazoles), with a trigonal component. The cupric ion would find itself in the same symmetry were it not for the three-fold set of tetragonal or rhombic distortions which trigonal complexes of this ion experience, believed to be due to a Jahn-Teller effect. The final geometry of a metal-insulin complex is the one which the metal prefers, and this will always be the case when the protein-contributed ligands are free to move.

At high NaCl (≥ 6%), a second form of zinc insulin crystallizes in which there are four metal ions per six Sanger units. One of these four zinc ions is on the trigonal axis and coordinated as in 2-Zn insulin crystals. The other three zinc ions are off the three-fold axis, and are tetrahedrally coordinated to two histidines (B5 and B10 from different Sanger units) and, probably, two water molecules (Bentley et al., 1976).

b) Carboxypeptidase (Hartsuck and Lipscomb, 1971)

In carboxypeptidase the zinc ion is bound to a histidine residue at position 69, a glutamic acid residue at position 72, and to another histidine at position 196. A water molecule (or hydroxyl ion, depending upon pH) is likely to be the fourth ligand, and the coordination is distorted tetrahedral. When substrate analogs are made to react with carboxypeptidase, several conformational changes occur in the protein which are large enough to be resolved by X-ray diffraction methods, and the zinc ion moves a detectable distance. These motions are indicative of flexibility within the molecule, including the coordination sphere. It is therefore not surprising that the zinc ion of carboxypeptidase can be displaced by many other metal ions [e.g. Cu(II), Hg(II)]. (There is considerable variation in enzymatic behavior among the several metal derivatives. Vallee and Coleman, 1964; Vallee et al., 1970.)

A situation different than that in insulin and carboxypeptidase would be one in which the amino acid residues involved in metal binding are constrained by virtue of their attachment to a single short segment of polypeptide or to separate segments of restricted flexibility. This may be the case in the intensely blue copper proteins (Chap. 3-3).

c) Myoglobin and Hemoglobin (Bolton and Perutz, 1970; Huber et al., 1970; Kendrew, 1963; Kendrew et al., 1960; Muirhead and Greer, 1970; Norvell et al., 1975; Padlan and Love, 1971; Perutz, 1970; Schoenborn, 1971; Stryer et al., 1964)

Four of the maximum of six coordinating atoms are fixed when iron is complexed with porphyrins. These are the pyrrole nitrogens, occupying positions defined as 1 through 4. Thus in hemeproteins, only one or two metal-protein links (at positions 5 and 6) can occur. In myoglobin, hemoglobin and erythrocruorin, there is an imidazole nitrogen of a histidine residue at position 5, and in position 6 a small ligand (e.g. water, oxygen, cyanide) or nothing. Apart from possible ruffling of the porphyrin ring, the geometrical factors which describe these complexes are the displacement of the iron ion from the plane of the pyrrole nitrogens, and the location and orientation with respect to the heme of the ligands in positions 5 and 6. In the aquo complex of ferrimyoglobin and ferrihemoglobin, the iron is 0.3-0.6 Å out of plane toward the coordinated imidazole nitrogen at position 5. The water molecule contributing an oxygen atom at position 6 is sandwiched between the heme and a distal histidine. This

$$\text{imidazole N} \xrightarrow{1.9\ \text{Å}} \text{Fe(III)} \xrightarrow{0.3\text{-}0.6\ \text{Å}} \text{porphyrin plane-X-imidazole}$$

situation is expected to hold in myoglobin and hemoglobin for all ligands X, such as water, fluoride, and acetate, which leave the five ferric valence electrons unpaired (high-spin case, weak field ligands). In ferrous deoxyhemoglobin, also a high-spin complex but five-coordinated rather than six, the displacement of the metal ion from the porphyrin plane is 0.7 to 0.8 Å toward the coordinated histidine, an increase over that for the high-spin ferric system. For complexes such as cyanoferrihemoglobin and oxyferrohemoglobin in which maximum spin-pairing occurs, leaving one (ferric) or no (ferrous) unpaired electron (low-spin case, strong field ligands), all coordination bonds are tightened compared with the high-spin case and the iron is anticipated to be in the plane of the pyrrole nitrogens. This has been found to be the case in lamprey ferrihemoglobin cyanide and the carbon monoxide complexes of myoglobin and erythrocruorin where the out-of-plane displacements are 0.1 Å or less. X-ray diffraction determined structures of oxymyoglobin and oxyhemoglobin are not available. These are low-spin complexes, and it can be assumed that the iron is in-plane. On this basis, the iron in the physiologically active ferrous proteins moves in and out of the plane upon oxygenation and deoxygenation. This movement is translated, possibly with slight amplification, to the rigidly linked histidine to which the iron is covalently bound, and is accompanied by conformational (tertiary structure) changes in the protein. In hemoglobin, which is a tetramer (four subunit chains) as compared with single chain myoglobin, there are also changes in quaternary structure which are likely to provide the mechanism for "cooperativity" (the affinity of a subunit for molecular oxygen depends upon the states of oxygenation of the other three subunits).

The porphyrin group extends over a relatively large area. In hemoglobin 60 atoms of the protein moiety are in van der Waals contact with the porphyrin ring.

d) Myohemerythrin and Hemerythrin

In analogy to the relation between hemoglobin and myoglobin, the oxygen-transporting polymeric protein hemerythrin and a monomeric myohemerythrin are found in the coelomic fluid and retractor muscles respectively of certain worms. Hemerythrin is an octamer of molecular weight 108,000 daltons, contains two iron atoms per subunit but no porphyrin, and binds one oxygen molecule

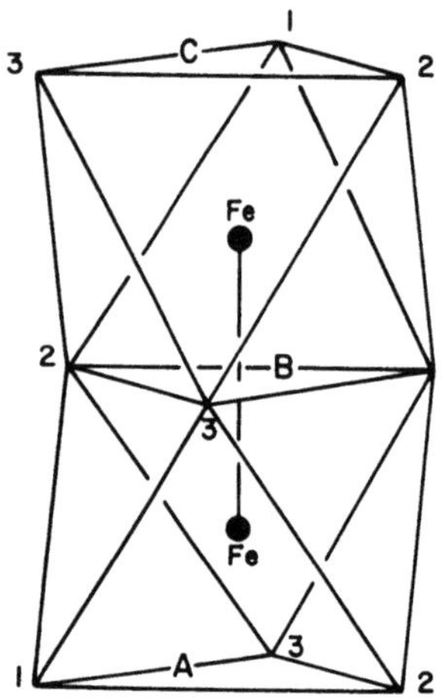

Fig. 2-2. Schematic representation of coordination polyhedra about iron atoms in subunit of hemerythrin (Stenkamp et al., 1976)

per two irons. The three-dimensional structure of methemerythrin has been interpreted from X-ray diffraction data (Stenkamp et al., 1976a,b; Ward et al., 1975). In accord with Mössbauer and magnetic susceptibility studies (Dawson et al., 1972; Moss et al., 1971; Okamura et al., 1969) which show them to be antiferromagnetically coupled, and spectroscopic studies which show them to be close together, the two irons of each monomer are found to be only 0.81 Å apart. A schematic of the structure of the iron complex in methemerythrin is given in Figure 2-2. Each iron is octahedrally coordinated. The Fe-Fe vector is along an approximate geometric three-fold axis. Curiously, at least in the methemerythrin from *Themiste dyscritum*, from subunit to subunit the Fe-Fe vectors are parallel. The two coordination octahedra of a subunit share a triangular face, B in Figure 2-2, at the corners of which are atoms which connect the two irons. The ligands at positions 1 and 2 of triangle B are believed to be contributed by glutamine and aspartic residues, respectively. The third vertex of this triangle does not appear to be associated with a protein side chain; at this position in methemerythrin water may be bound, and in the physiological state, molecular oxygen. As is the case with opposing faces of an octahedron, triangles A and C are rotated with respect to triangle B by 60° around the trigonal axis. Positions 1 and 2 of triangle A and all positions of C appear to be associated with histidine residues, and position 3 of A with a tyrosine residue.

Myohemerythrin, a monomer of 13,900 daltons, contains two iron atoms, binds one oxygen molecule, and is thought on the basis of chemical and physical properties to be structurally similar to the subunit of hemerythrin. The three-dimensional structure of myohemerythrin, in the form of the ferric azide complex, at 5.5 Å resolution (Hendrickson et al., 1975) shows this to be the case, both with regard to iron coordination and polypeptide chain configuration. In the monomer and the subunit, the two-iron center appears as though held inside four roughly parallel rods, which are approximately equal lengths of helical polypeptide joined together at the ends (Fig. 2-3). Access to the iron

Fig. 2-3. Drawing of electron density in myohemerythrin (Hendrikson et al., 1975)

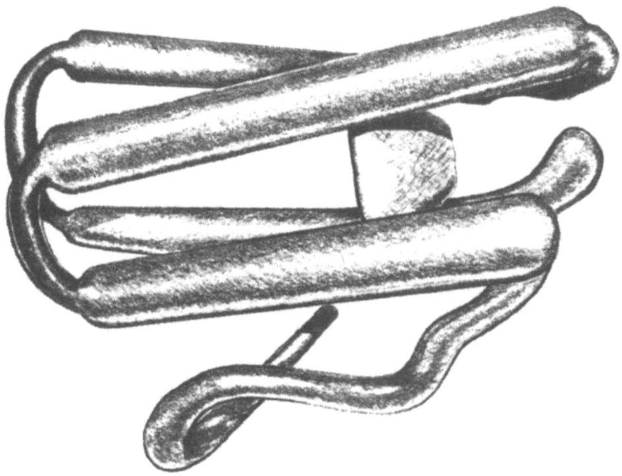

atoms is restricted. Chemical and X-ray data taken together suggest that the residues in the neighborhood of the iron center (apart from the ligands themselves) are in general nonpolar.

e) Cytochrome c (Dickerson et al., 1971; Dickerson and Timkovich, 1975; Salemme et al., 1973)

Essentially identical iron coordination is found in the crystal structures of cytochrome c from tuna (12,200 daltons), c_2 from *Rhodospirillum rubrum* (12,500 daltons) and c_{550} from *Paracoccus denitrificans* (14,900 daltons). As in myoglobin and hemoglobin, the iron of cytochrome c is complexed to protoporphyrin IX and to a histidine side chain in position 5. However, the bonding of the heme in cytochrome c has several distinctive features. Methionine sulfur is coordinated to the iron at position 6. Between the porphyrin and the protein chain there are two covalent links, which can be viewed as arising from the reduction of the vinyl groups by cysteine residues. The latter cysteines are separated by two other amino acids in the primary sequence. One of the porphyrin propionic groups is hydrogen bonded to amino acid residues in the interior of the protein. The other propriionic group, also apparently held in place by hydrogen bonds, is half-exposed to the solvent phase. The polypeptide chain weaves around three edges of the heme, forming a hydrophobic crevice and leaving the fourth edge exposed. Positively charged lysine side chain amino groups are situated near the ends of the exposed edge of the heme.

At neutral pH, the ferric ion in the oxidized enzyme is low spin, which indicates substantial bonding to both the methionine sulfur and the imidazole nitrogen (Boeri et al., 1953). As evidenced by several physical and chemical properties (e.g. chromatographic elution, proteinase digestion, solubility; Okunuki, 1966), and seen in X-ray diffraction-determined structural differences, the protein moiety of cytochrome c undergoes significant conformational changes when the valence of the iron changes.

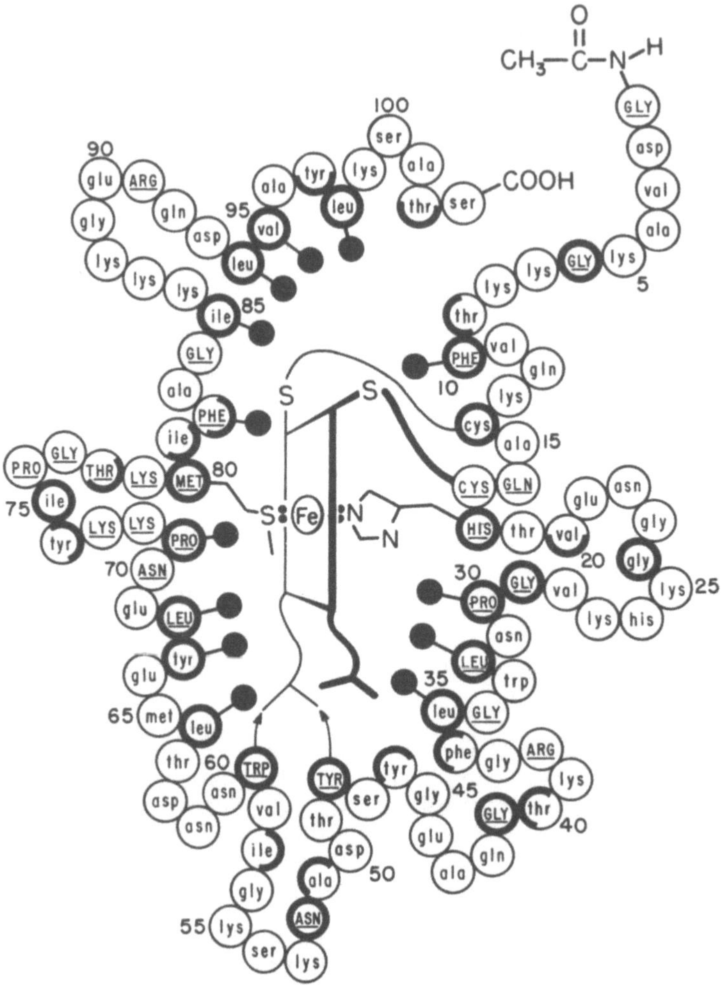

Fig. 2-4. Diagram of reduced cytochrome c (tuna). *Black dots*: side chains packed against the heme; *thin circles*: side chains on the outside of the protein; *heavy circles*: buried side chains. Those residues which are *underlined* are evolutionarily invariant. The exposed edge of the heme group is indicated by a *bold line* (Dickerson and Timkovich, 1975)

In the reduced state, the variation in environment around the heme is more sharply defined; this is schematically shown in Figure 2-4. There is usually tyrosine or phenylalanine at "position 74", and a mechanism involving aromatic side chains has been postulated for the transfer of electrons on the "left side" of the molecule. However, in cytochrome c_{550} there is a leucine at the equivalent position, the non-aromaticity of which is inconsistent with the latter hypothesis.

f) Cytochrome b_5 (Mathews et al., 1971a)

Cytochrome b_5 solubilized by detergent has a molecular weight of about 16,000 daltons. Solubilization with pancreatic lipase is accompanied by the cleavage of about 40 amino acid residues from the amino-terminus and reduction of the molecular weight to 11,000. Crystals of the latter molecule have been studied by X-ray diffraction and the three-dimensional structure deduced. As in myoglobin, hemoglobin, and cytochrome c, the iron is coordinated to the four pyrrole nitrogens of protoporphyrin IX, but instead of being linked to one imidazole nitrogen, it is linked to two. Thus, in cytochrome b_5, the iron is directly coordinated to six nitrogens. The heme group of cytochrome b_5, like that of myoglobin and hemoglobin and unlike that of cytochrome c, is not covalently bound to the protein. It adheres to a hydrophobic groove with the two vinyl groups directed toward the interior of the protein and the two propionic groups toward the exterior aqueous environment. One propionic group is bonded to the surface of the protein and the other projects into the surrounding solution. The iron-linked histidines, 39 and 63, appear to be constrained by both hydrogen and van der Waals bonds to other residues so that small molecules or ions are hindered from complexing the metal ion. That is, in order for a histidine to move away from the iron so as to permit a change in ligand, conformational change in the protein appears to be necessary. Initial X-ray study of the effect of reduction of the heme iron of cytochrome b_5 from ferric to ferrous indicates that any changes in protein conformation which occur are likely to be small (Mathews et al., 1971b).

g) Copper, Zinc Superoxide Dismutase

The X-ray structure of bovine erythrocyte Cu(II), Zn(II) superoxide dismutase at 3 Å resolution provides the essential features of the course of the polypeptide backbone and the coordination of the two metal ions (Richardson et al., 1975). The subunits of this dimer have a molecular weight of 16,000 daltons, are essentially identical at 3 Å resolution, and each contains one cupric and one zinc ion. The large contact area between subunits is mainly hydrophobic and does not include a covalent linkage. (There is, however, a disulfide bond within each subunit.) The active sites of the dimer are well separated, the distance between the two coppers being 34 Å. Within a subunit, the Cu and Zn are about 6 Å apart. Almost half of the polypeptide chain belongs to an eight-stranded "barrel" of antiparallel β-pleated sheet. There are two long hydrophilic loops coming from an area of the barrel close to one edge (Fig. 2-5). The Cu and Zn sites are located between the loops and the edge such that three of the Zn ligands come from side chains on one of the loops and the fourth from a nearby barrel residue, and conversely for the Cu ligands. Copper shares its single loop ligand, from a histidine, with the Zn.

The Zn coordination geometry is approximately tetrahedral, with ligands from three histidine and one aspartic acid side chains.

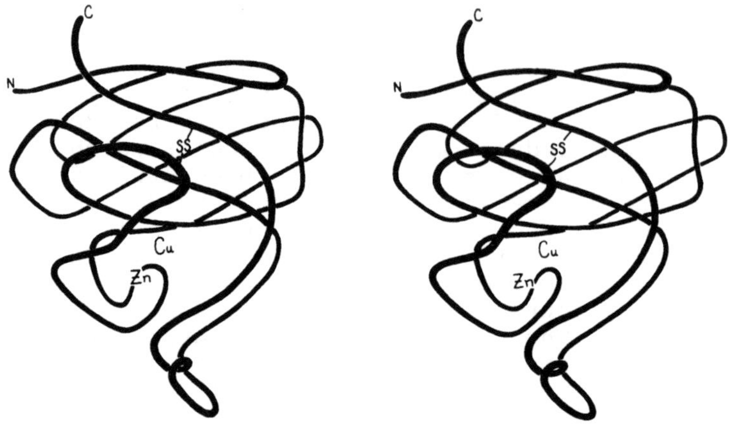

Fig. 2-5. Simplified schematic drawing in stereo which shows the polypeptide backbone and locations of the Cu and Zn ions of a superoxide dimutase subunit. The single disulfide bridge is also indicated (Richardson et al., 1975)

There are four histidine-contributed ligands to the cupric ion, disposed in square planar symmetry. The geometrical structures of these Cu and Zn sites are typical of many small complexes of the same metals, and, with the sharing of a histidine side chain, bear a particularly close resemblance to imidazole complexes (Freeman, 1967). There is ready access to the Cu from the solvent phase in one of the two directions normal to the coordination plane.

h) Rubredoxin

Rubredoxins differ from other iron-sulfur proteins in not having acid-labile sulfur. On the contrary, a single iron is bound to four cysteine sulfurs in a complex of approximately tetrahedral geometry, and no other residues or non-protein ligands are involved, as shown in Figure 2-6 (Watenpaugh et al., 1973). There can be two sites per molecule, each with one iron, as in the rubredoxin from *Pseudomonas oleovorans* (single polypeptide chain, 19,500 daltons), but in general the rubredoxins are smaller and contain a single active center (e.g. from *Clostridium pasteurianum*, 6127 daltons). The high resolution X-ray structure is from ferric rubredoxin; the ferrous coordination seen at lower resolution also appears to be tetrahedral.

i) Ferredoxin and High Potential Iron-Sulfur Protein (HiPiP)

Ferredoxins are those iron-sulfur proteins with equal numbers of iron and labile sulfur atoms which are reducing (negative midpoint redox potentials at pH 7). There are ferredoxins with 2Fe-2S centers from both prokaryotic and eukaryotic sources, and ferredoxins with 4Fe-4S centers which are mainly bacterial in origin. In the absence of a structure determined from X-ray diffraction data, knowledge of the coordination of the metal in

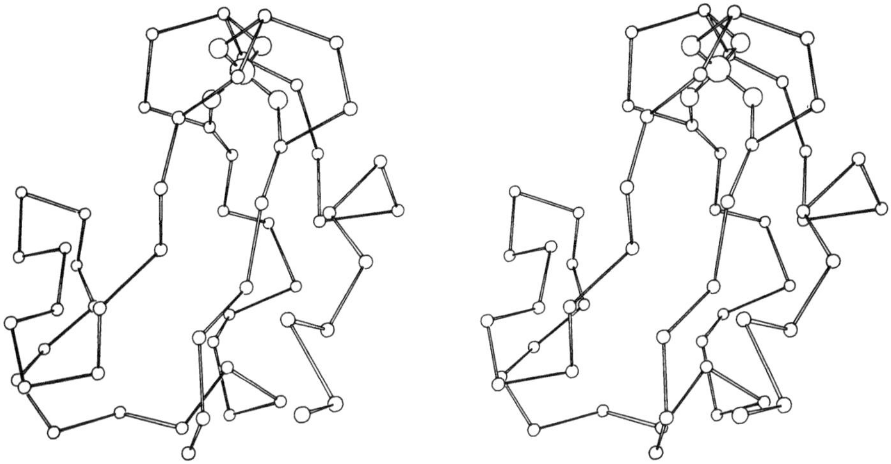

Fig. 2-6. Plot in stereo of the C^{α} atoms and iron-sulfur complex of rubredoxin. The largest circle is the iron and the neighboring four circles, the sulfurs. (Watenpaugh et al., 1973).

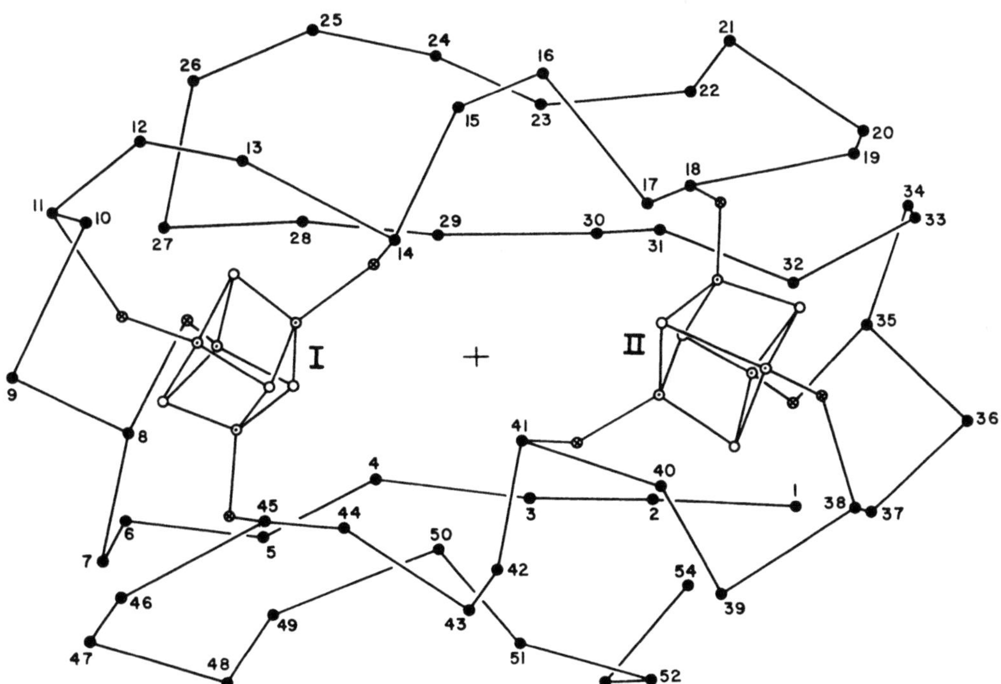

Fig. 2-7. Plot of the α carbon ●; cysteine sulfur ⊗; iron ⊙; and inorganic sulfur O, positions, showing approximate two-fold symmetry in bacterial ferredoxin (Adman et al., 1973)

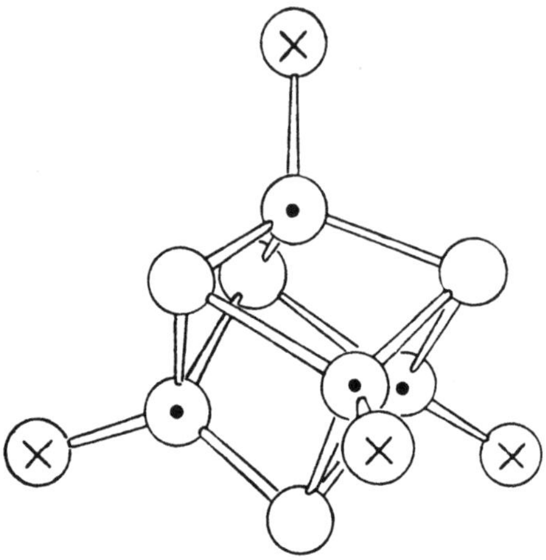

Fig. 2-8. Schematic of the iron-sulfur cluster in bacterial ferredoxin and HiPiP. Iron ⊙; inorganic sulfur O; cysteine sulfur ⊗

the 2Fe-2S proteins comes from chemical and spectroscopic studies (Chap. 5-2). The three-dimensional X-ray picture of the ferredoxin from *Peptococcus aerogenes* shows two 4Fe-4S complexes 12 Å apart (Fig. 2-7) (Adman et al., 1973). The eight atoms of each center are located at the corners of an approximate cube, with each Fe(S) having three S(Fe) as nearest neighbors; this "cubane" structure is completed by a link between each iron atom and a sulfur atom of a cysteine residue (Fig. 2-8). Thus the coordination around the irons is an approximate tetrahedron of sulfur atoms which differs from that of rubredoxin in that three of the sulfurs are inorganic and only one is mercaptide.

High potential iron-sulfur proteins also contain the cubanelike cluster of Figure 2-8 (Carter et al., 1971), but, by definition, are oxidizing at pH 7. HiPiP is paramagnetic and gives an EPR signal only in the oxidized form [valence model 3Fe(III), 1Fe(II)]; whereas those ferredoxins (reducing at pH 7) which have the 4Fe-4S clusters are paramagnetic and EPR-active only when reduced [valence model 1Fe(III), 3Fe(II)]. [The "intermediate form" - 2Fe(III), 2Fe(II) - is diamagnetic.] In view of the great similarity of the metal-sulfur centers in these two groups of proteins, the differences in physical properties are ascribed to effects of the protein-contributed environment (Carter et al., 1974).

3. Covalency

Covalency of bonding is related to the extent to which metal electrons are shared by the ligands. In terms of a molecular orbital description (Chap. 6-4), the electrons occupy spatial orbitals ψ which are combinations of metal orbitals ψ_M and ligand orbitals ψ_L, given to first approximation by

$$\psi = \alpha\psi_M - \alpha'\psi_L$$

In such a molecular orbital, the fractional occupancy of the metal is α^2, of the ligands is $(\alpha')^2$, and there is an "overlap population" (of the order of 0 to 0.1). Thus $\alpha^2 \approx 1$, $(\alpha')^2 \approx 0$, reflects a wholly ionic situation, and $\alpha^2 \approx (\alpha')^2 \approx 0.55$, a wholly covalent one. A quantitative measure of electron delocalization is available when ligand hyperfine structure appears in EPR spectra and/or when ENDOR can be observed (Chaps. 1-6, 6-7). This kind of data forms the basis for the covalency entries of Table 2-3 wherein the nitrogen lonepairs involved in molecular orbital formation are taken to be in trigonal hybrid orbitals. There are not marked differences in delocalization among the cases listed, and related "model compounds" yield values in the same range.

4. Some Aspects of Differences in Heme Binding

In hemoglobin and myoglobin, the protein moieties do not contribute an atom to coordination position 6, while in cytochromes c and b_5 respectively methionine sulfur and imidazole nitrogen atoms are found in this position. The limited accessibility of the iron in cytochrome c was noted in early studies of the reactivity of this protein. Compared with the corresponding hemoglobin and myoglobin derivatives, ferricytochrome c reacts more slowly with fluoride and cyanide, and ferrocytochrome c more slowly with carbon monoxide. The cyanide reactions are thermodynamically characterized by favorable enthalpy change and unfavorable entropy change for ferrimyoglobin, and conversely for ferricytochrome c (George, 1956). Furthermore, the optical absorption spectrum of ferrocytochrome c is much like that of a "hemochromogen" (coordination complex of ferrous heme with nitrogenous bases in positions 5 and 6). Thus, several lines of evidence pointed to the relative inaccessibility of the heme iron in cytochrome c in advance of the appearance of the X-ray determined structure. More recent experiments show that the heme group by itself, even if it were not shielded by the protein, cannot account for the "active site" properties of this enzyme. By means of site-specific antibodies, Smith et al. (1973) have demonstrated separate oxidase and reductase reaction sites on cytochrome c. The complex kinetic curve of the reduction of ferricytochrome c by hydrogen atoms suggests that intraprotein radical transfer proceeds from several primary reaction sites on the outer surface of the enzyme (Shafferman and Stein, 1974). These and other recent results, together with earlier ones (see Chap. 1-7), implicate the protein and porphyrin moieties of cytochrome c in electron transfer, and leave for the iron in this molecule the role of storing an oxidizing or reducing equivalent.

Chapter 3
Copper

Taken together the following reviews cover most topics in the biochemistry and biophysics of copper: Brill et al., 1964; Freeman, 1967; Frieden, 1974; Frieden et al., 1965; Malkin and Malmström, 1970; O'Dell and Campbell, 1970; Peisach et al., 1966; Vänngård, 1972.

1. Cupric Peptides

Cupric peptides are as close chemically to cupric proteins as is possible for small model complexes to be. A very large number of coordination possibilities exist for the peptides, depending upon their length, whether or not the side chains have coordinating groups and what these may be, and what the pH is. Of course amino acids themselves complex the cupric ion. (Only data from L-amino acids and peptides containing L-amino acids will be referred to in this discussion.) Tsangaris and Martin (1970) have shown that the 2:1 complexes of many amino acids (e.g. alanine, glutamic acid, valine, tyrosine) at a pH such that the α-amino group is deprotonated exhibit absorption maxima at about 620 nm with an absorptivity of about 50 $M^{-1}cm^{-1}$, and negative extrema in the CD spectra at about the same wavelength with ellipticities averaging -800 deg cm^2/decimol. (2:1 imino acids show stronger absorption and positive ellipticity in this region.) In these complexes, the cupric ion is coordinated to two negatively charged carboxyl oxygens and to two amino (or imino) nitrogens. 1:1 cupric dipeptide complexes at a pH such that the amide (peptide) hydrogen is ionized are coordinated through one amino nitrogen, one amide nitrogen, one carboxyl oxygen, and one water or OH^- oxygen. The maximum absorptivity for these complexes is about 80 $M^{-1}cm^{-1}$ near 640 nm, and the corresponding ellipticities average -1700 deg cm^2/decimol. Tri- and tetra-peptide 1:1 complexes with the cupric ion, at alkaline pH, involve two or three deprotonated amide nitrogens, one amino nitrogen, and one or zero carboxyl or water or OH^- oxygens. The wavelengths of the peak absorptivity are now blue-shifted to about 550 nm and the maxima are significantly greater, about 150 $M^{-1}cm^{-1}$; the extrema in ellipticity average -3300 in this region, the range being large.

Histidine-containing peptides are of special interest because the imidazole group is so widely involved in metal binding. As with the peptides of the preceding paragraph, cupric ion complexes with histidine-containing peptides show movement of the absorption maximum to shorter wavelengths and an intensification

Table 3-1. Visible absorption bands of the 1:1 complexes of Cu(II) with glycylgycylglycyl-L-histidine

$Cu(II)L^+$		$Cu(II)L$		$Cu(II)L^-$		$Cu(II)L^{2-}$	
λ_{max} nm	a_{max} $M^{-1}cm^{-1}$	λ_{max} nm	a_{max} $M^{-1}cm^{-1}$	λ_{max} nm	a_{max} $M^{-1}cm^{-1}$	λ_{max} nm	a_{max} $M^{-1}cm^{-1}$
715	40	660	63	590	65	550	112

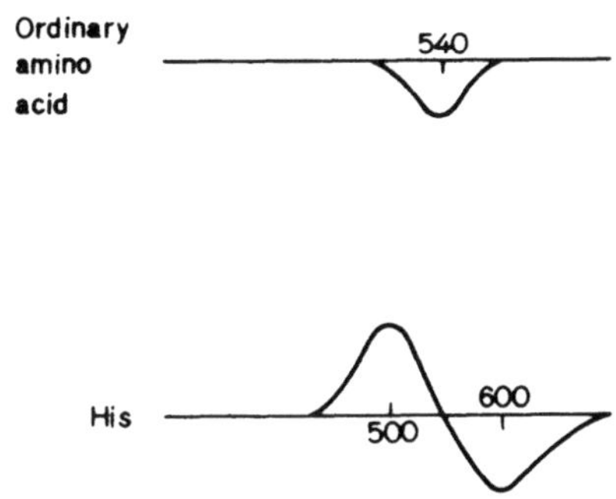

Fig. 3-1. CD bands of cupric complexes with amino acids and peptides (Martin, 1974)

of absorption as the pH is raised, reflecting coordination to more nitrogens and increasing constraint upon the geometry. This behavior is demonstrated by the data in Table 3-1 from the 1:1 complexes of Cu(II) with glycylglycylglycyl-L-histidine (Bryce and Gurd, 1966). Here CuL^+ refers to the complex in which the metal ion is bound to imidazole and α-amino nitrogens; the carboxylate group is also involved and the net charge is +1. As the pH is raised, at first one peptide proton ionizes and the copper links to the nitrogen [Cu(II)L]; then a second peptide nitrogen is deprotonated and coordinates to the metal $[Cu(II)L^-]$; and, finally, the third amide ionization and coordination occurs $[Cu(II)L^{2-}]$. In the final complex, the "in-plane" ligands are one imidazole and three peptide nitrogens, and the α-amino nitrogen is probably in an apical position. The progression in energy and intensity is monotonic with pH for tri- and higher peptides of this kind.

While most L-amino acids and peptides with L-amino acid residues in complexes with the cupric ion yield only negative CD bands in the visible region, histidine-containing cupric peptides exhibit both + and - bands, as indicated in Figure 3-1.

Cupric peptides have a positive CD band at 300-315 nm, but cupric amino acid complexes do not (Martin, 1974). Although there is no corresponding absorption maximum, a chromophoric function for the amide nitrogen (necessarily deprotonated and linked to the metal ion) is nonetheless implicated by this data. The ionized peptide nitrogen is a strong electron donor in the sense of coordinate-covalency and may also be an electron-transfer donor to Cu(II).

Table 3-2. Cupric glycyl peptides; peptide nitrogens ionized and linked to the metal ion

Complex	λ_{max}	a_{max}	Ligand amino nitrogens	Ligand amide nitrogens
Diglycine (2:1)	640 nm	84	1	1
Triglycine (1:1)	560	162	1	2
Tetraglycine (1:1)	520	150	1	3
Tetraglycylglycine (1:1)	510	147	1	3

The absorptivity of peptide complexes of the type $Cu(II)L^+$ is not minimal. For example, copper sulfate and cupric insulin have distinctly smaller d-d absorptivities, about 10 $M^{-1}cm^{-1}$. The latter feeble intensity is unlikely to indicate the presence of an odd symmetry component which is a property of the environment of the metal ion, but rather that some vibrational mode of the complex results in the metal ion spending part of its time away from a center of symmetry (vibronic admixture of odd states). The cupric ammonia complex, $Cu(NH_3)_4^{2+}$, acquires its deep blue color from the same vibronic mechanism, but is several times more intense ($a^{600\ nm}$ = 50 $M^{-1}cm^{-1}$) because of the greater copper-ligand coupling (stronger coordination). In general, increasing structural complexity and decreasing symmetry is accompanied by increasing absorptivity. At first this may be due, as in the case of amino acid and dipeptide complexes, to groups (residues) outside the immediate coordination square or rectangle being located in such a way as to produce a small potential component at the metal ion which admixes to the d-orbitals ones of odd character. The structures of many cupric peptides are known from the X-ray diffraction studies of Freeman (1967) and his collaborators. Except in those crystals where the copper atom lies at a symmetry center, the four liganded atoms deviate from their plane of best fit, in a flattened tetrahedral arrangement, and the metal ion is displaced from the plane. Table 3-2 demonstrates that distortions of this kind, and concomitant ones in the disposition of the other atoms of the chelate rings, are sufficient to produce moderate absorptivities in peptides which do not have side-chains. In some cupric peptides, the liganded atoms lie as much as 0.16 Å away from the average coordination plane; the deviation of the copper atoms, as much as 0.25 Å. The ligand field and the fields of atoms and groups further away can be additive or subtractive in their effects on light absorption.

Other cupric complexes which are known to be tetrahedrally distorted have moderately intense bands in the visible or near infra-red, e.g. tetrachlorocuprate has a molar absorptivity of about 120 cm^{-1} at 1050 nm. Stronger bands at half the wavelength (520 nm) are exhibited by penicillamine complexes of the cupric ion (Wilson and Martin, 1971), but these are now known to be multi-metal, multi-sulfur compounds (Schugar et al., 1976).

Certain generalizations from the X-ray determined structures of cupric peptides are especially worth nothing. As compared with free peptides, the length of the C=O bond is increased and that of the C-N bond shortened (evidence of increased double-bond character) when the proton on an amide nitrogen is replaced by a cupric ion. It is well-known that there is double-bond character in the C-N bond of free peptides, with the consequence that the atoms C_α-CO-NH of the peptide group are coplanar. That this coplanarity is preserved in ionized peptide groups linked to Cu(II) is consistent with the preceding observation. The Cu-amide N bond lengths average 1.92 Å and the disposition of the atoms about the coordinated amide nitrogens is closely trigonal. Cu-amino N and Cu-imidazole N bond lengths average 2.00 Å, and Cu-O bond lengths are 1.97-1.98 Å. Correlation of the structure of cupric peptides in the solid state with absorption measurements on the same complexes in solution confirms that ligand field differences (the energies of the d-d transitions) increase as oxygens are replaced by nitrogens in the coordination plane. One also finds that the coordination number decreases steadily from 6 to 4 as the closest ligand atoms are changed one at a time from four oxygens to four nitrogens - that is, the ability of the cupric ion to bond axial ligands (distant 2.30-2.96 Å) decreases as the number of in-plane nitrogens increases.

2. EPR of Cupric Peptides and Related Complexes

A pioneering and broad look at the EPR properties of the cupric ion in proteins and model systems was taken by Mälmstrom and Vänngård (1960) who discovered the much reduced low-field hyperfine splitting in the blue proteins. Later this chapter the latter phenomenon and related topics will be discussed; small model complexes are the present subject. Gould and Mason (1966), utilizing EPR spectroscopy, showed that the structure of the complex which is formed when the cupric ion reacts with glycylglycine is pH dependent. They inferred from the five ligand hyperfine lines observed in the complex at pH 10 that the metal ion was binding to the (deprotonated) amide nitrogen as well as the amino nitrogen, in support of the conclusions from titrimetric, spectrophotometric, and X-ray studies outlined in the preceding section.

Bryce (1966) investigated the EPR of cupric complexes of histidine-containing peptides and used the data to arrive at molecular orbitals which describe the coordination. In all of the complexes of this study, the metal ion is bound to four nitrogens in a distorted square-planar array. Delocalization of the cupric hole onto these nitrogens is explicitly demonstrated by the presence of ligand superhyperfine splittings in the EPR spectra. A spectrum taken at room temperature is shown in Figure 3-2 and at 102 K, in Figure 3-3. $g_{||}$ and $A_{||}$ (Chap. 6-6, 7) can be measured from the frozen solution spectra, but $g_\perp$ and $A_\perp$ can not. Computer simulation of the low temperature EPR spectra gives reliable g and A values, thereby providing $g_\perp$ and $A_\perp$ and a check on $g_{||}$ and $A_{||}$. Subject to the limitations pointed out by Falle and Luckhurst (1965), there is an independent means of arriving

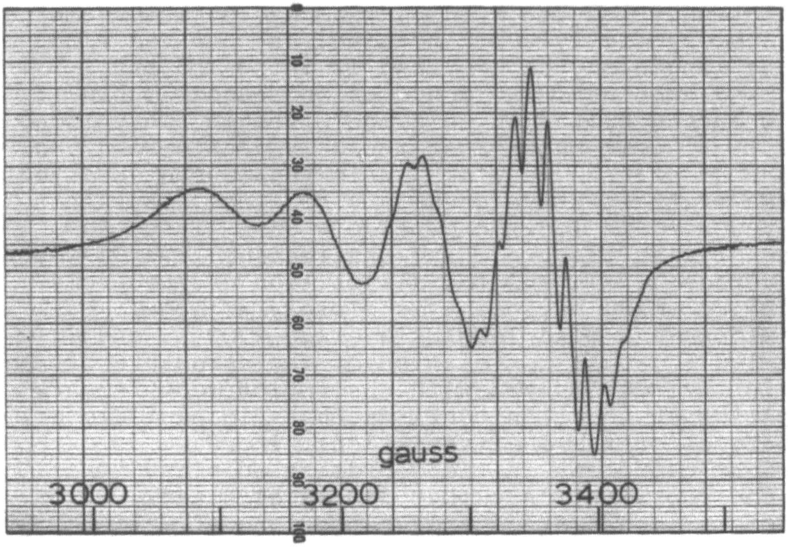

Fig. 3-2. EPR spectrum of the cupric complex of acetylglycylglycyl-L-histidylglycine in aqueous solution, pH 11, at room temperature (Bryce, 1966)

at $g_\perp$ and $A_\perp$ through the combination of low and room temperature data. For rapidly tumbling small complexes in solution, the following relations hold, approximately:

$$\langle g \rangle_{\text{solution}} \cong \frac{1}{3}(g_x + g_y + g_z)$$

$$= \frac{1}{3}(g_{||} + 2g_\perp) \text{ for axial solutions,}$$

$$\langle A \rangle_{\text{solution}} \cong \frac{1}{3}(A_x + A_y + A_z) = \frac{1}{3}(A_{||} + 2A_\perp),$$

$$\langle A^{\text{Ligand}} \rangle \equiv \langle A^L \rangle \cong \frac{1}{3}(A^L_{||\text{to bond}} + 2A^L_{\perp\text{to bond}}) = A_{\text{Fermi}}$$

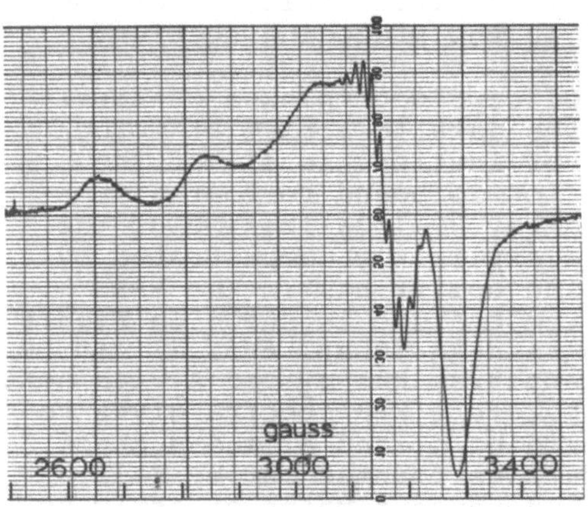

Fig. 3-3. EPR spectrum of the cupric complex of Figure 3-2 at 102 K (Bryce, 1966)

It follows then that $g_{\perp} \cong \frac{1}{2}(3\langle g\rangle_{solution} - g_{||})$, and $A_{\perp} \cong \frac{1}{2}(3\langle A\rangle_{solution} - A_{||})$.

Bryce used the molecular orbital model (Chap. 6-4 and Table 6-2)

$$\psi_{ground} = \alpha d_{x^2-y^2} - \frac{\alpha'}{2}(-\sigma_x^{(1)} + \sigma_y^{(2)} + \sigma_x^{(3)} - \sigma_y^{(4)})$$

$$\psi_{\pi}^{in\text{-}plane} = \beta d_{xy} - \frac{1}{2}(1 - \beta^2)^{1/2}(p_y^{(1)} + p_x^{(2)} - p_y^{(3)} - p_x^{(4)})$$

$$\psi_{\pi xz}^{out\text{-}of\text{-}plane} = \delta d_{xz} - (1 - \delta^2)^{1/2}(p_z^{(1)} - p_z^{(3)})/\sqrt{2}$$

$$\psi_{\pi yz}^{out\text{-}of\text{-}plane} = \delta d_{yz} - (1 - \delta^2)^{1/2}(p_z^{(2)} - p_z^{(4)})/\sqrt{2}$$

to formulate expressions for the spin-Hamiltonian parameters $g_{||}$, $g_{\perp}$, $A_{||}$, and $A_{\perp}$. Here $\sigma(\frac{i}{i+2})$ is the hybrid orbital $\sqrt{\frac{h}{h+1}}\,p(\frac{i}{i+2}) \mp \sqrt{\frac{1}{h+1}}\,s(\frac{i}{i+2})$. (Note that the molecular orbital involving d_{z^2} does not enter this calculation.) For example,

$$A_{||} = -\kappa P\alpha^2 - P[2.0023 - g_{||} + \frac{4}{7}\alpha^2 + \text{smaller terms}].$$

Various calculations of P for Cu(II) (actually ^{63}Cu and ^{65}Cu will have slightly differing P values through differences in the nuclear g-value) place it in the range 0.036 to 0.040 cm^{-1}. κ is found to be close to 0.43 for most cupric complexes. Prior optical studies provided the ligand field transition energies which appear in the formulae. The EPR data were matched when the molecular orbital coefficients were within well-defined ranges. The squares are given below (α' is obtained with the help of the normalization relation $\alpha^2 + (\alpha')^2 - 2\alpha\alpha'S = 1$ given in Chapter 6-4); σ-bonding: $0.81 \le \alpha^2 \le 0.82$; $0.27 \le (\alpha')^2 \le 0.29$; π-bonding: in-plane, $0.64 \le \beta^2 \le 0.74$; out-of-plane, $0.65 \le \delta^2 \le 0.76$. The ligand admixtures imply moderate covalency in the σ-bonds and stronger covalency in the π-bonds.

An independent estimate of $(\alpha')^2$ is available through the ligand nuclear-hyperfine splittings (Chap. 6-7). Since the nitrogens involved in metal-binding in these complexes are almost all peptide and imidazole, the hybridization was taken as trigonal; hence

$$(2\alpha_N)^2 = (\alpha')^2 = 9A_F^N/4\pi g_N\beta\beta_N|2s^N(0)|^2$$

The values of $(\alpha')^2$ obtained by the two methods do not differ by more than 0.02. There can, of course, be some variation in delocalization, both σ and π, among the nitrogens.

There have been many EPR investigations of cupric complexes in which methods of analysis similar to those just described have been used. The results are summarized in Table 3-3. Delocaliza-

Table 3-3. Bonding parameters of cupric complexes, derived from EPR

Complex	Ligand atoms	α^2	$(\alpha')^2$	β^2	δ^2	References
Bis-salicylaldehyde-imine	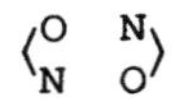	0.83	0.26	0.72	0.91	Maki and McGarvey (1958)
Benzoyl hydrazone	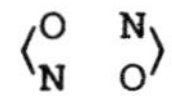	0.82	0.27	0.86	0.62	Hsu (1971)
o-Vanillin-ethylene diamine	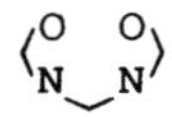	0.80	0.29	0.79	0.77	Hsu (1971)
Peptides containing a histidine	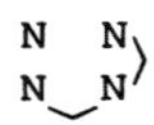 	0.81- 0.82	0.26- 0.29	0.64- 0.74	0.65- 0.76	Bryce (1966)
Etioporphyrin II	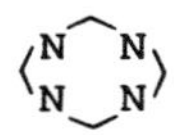	0.79	0.33	0.92	0.90	Roberts and Koski (1960)
Protoporphyrin IX dimethyl ester	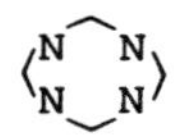	0.75	0.35	~ 1	~ 1	Hsu (1971)
Phthalocyanin	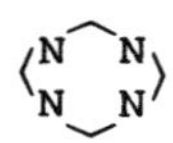	0.74	0.37	~ 1	~ 0.6	Guzy et al. (1969)
Tetraphenylporphin	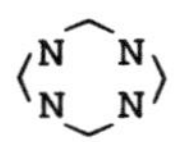	0.74	0.36	~ 1	~ 1	Assour (1965)

Fig. 3-4A and B. EPR spectrum of *Pseudomonas fluorescens* azurin (X-band, 77 K) (Brill et al., 1968). The low field region shown in (B) was taken at 16× greater gain than the record of (A)

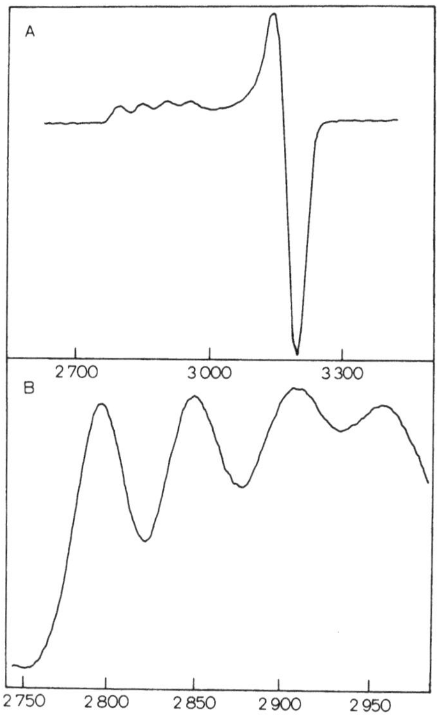

tion into σ-bonds increases as more of the liganded atoms become part of the same chelate ring, and there appears to be a concomitant decrease of in-plane π-bonding. It is interesting to note the differences in σ-bond delocalization among the last four compounds, which have closely related ring systems.

Rather different in appearance than the EPR spectra just discussed are those from "Type 1" cupric ions in blue proteins, to be described in the following section. An example is given in Figure 3-4, from azurin at 77 K. Note the much reduced separation of the low field copper hyperfine bands and the distortions of their pattern.

3. The Blue Proteins

There is a group of copper-containing proteins, the "blue proteins", with a unique set of optical and magnetic properties, including the intense blue color from which the group derives its name. Some of these have several coppers (e.g. ceruloplasmin, or ferroxidase, has seven or eight; laccases, four), plastocyanin has two; and several have a single copper (e.g. the azurins, from microorganisms; stellacyanin, from the lac tree; umecyanin, from horseradish root). The molecular weight per copper is as small as 11,000 daltons for plastocyanin and as large as 30,000 daltons for plant laccase, but usually is 14,000-20,000. The marine invertebrate oxygen-transporting pigment hemocyanin is blue too, but is not considered one of *the* blue proteins because the molar absorptivity per copper is low by comparison and because there remains some question about the valence of the metal ions in the chromophores. The true blue proteins are colored when cupric and bleached when cuprous.

For the present, the discussion will emphasize the blue proteins which have a single copper: these exhibit all the characteristic optical and magnetic properties (Mason, 1963), eliminate the need for sorting out contributions from "nonblue" cupric sites, and remove from consideration models which invoke copper-copper interactions. In spite of the similarities of the cupric sites, these macromolecules can differ significantly. For example, stellacyanin is 60% protein, 20% carbohydrate, and 20% hexose-amine (Peisach et al., 1967); umecyanin is a few per cent carbohydrate and contains three glucosamine residues per molecule (Stigbrand, 1971); while the azurins are, but for the copper, all protein (Ambler and Brown, 1967). The amino acid sequences of azurins from different microorganisms have much in common (Ambler, 1968).

EPR, optical absorption, and CD spectra for azurin and stellacyanin are shown in Figure 3-5. These are not the experimentally measured spectra but are computed from the experimentally determined magnetic and optical parameters with the assumption of Gaussian lineshapes, thereby permitting a lineshape-free comparison with spectra from models.

Experimental values of quantities measured on blue proteins containing a single cupric ion are listed in Table 3-4. Note that, apart from ligand hyperfine interactions, frequencies (wavenumbers[2]) at which absorptivity maxima occur and the corresponding values of bandwidth, there are 16 pieces of magnetic and optical data to describe the metal binding site. Rarely is one able to characterize the electronic structure of a single atom in a molecule so broadly, a situation which one wishes to exploit, for this kind of cupric coordination site appears to be uniquely provided by proteins. The salient experimental features are the relatively high intensities of the absorption bands, the strong circular dichroism associated with them, and the collapsed low field copper hyperfine splittings. The maximum absorptivities of the band at ~16 kK are two orders of magnitude greater than those of the simplest complexes (e.g. sulfate or amino acid), and more than one order of magnitude greater than those from most peptides. Even the less intense bands of the blue proteins at ~13 kK and ~19 kK are significantly stronger than the most intense ones from small cupric complexes. Copper complexes with aldehydes of oxalyldihydrazone have absorptivities of the order of $10^4 M^{-1} cm^{-1}$ at about 17 kK (Brill et al., 1964; Osaki, 1966), but these complexes show no EPR signal and are likely to be cuprous with strong metal to ligand charge transfer states accounting for the intense visible bands. At neutral pH, the hyperfine splitting $|A_z|$ for these blue proteins is in the range 30 to $60 \times 10^{-4} cm^{-1}$ while the cupric ion in non-"blue" protein sites and in small molecule complexes has $|A_z|$ values ranging from 140 to $220 \times 10^{-4} cm^{-1}$.

---

[2]The symbol K, for Kayser = 1 wavenumber = 1 cm^{-1}, will be used in Chapter 4; e.g. 1 kK = $10^3 cm^{-1}$.

Table 3-4. Magnetic and optical properties of blue proteins containing a single cupric ion

(a) Spin-Hamiltonian parameters

Protein	MW[a]	pH	g_x	g_y	g_z	$\|A_x\|$[b]	$\|A_y\|$[b]	$\|A_z\|$[b]	Notes, refs.
azurin *Ps. aerug.*	14,000 *(fluorescens)*	neutral[c]	2.056	2.039	2.264	< 7	< 7	56	1,2,3
Stellacyanin	20,000[d]	1-9	2.079	2.026	2.289	~29	56	~36	[e]3,5,6
		10-11.5	2.089	2.025	2.312	NE[f]	73	<17	5
		12.8	2.04	2.04	2.16	NE	NE	196	5
Umecyanin	14,600	neutral	2.05	2.05	2.317	NE	NE	35	7,8,9
		9.5	~2.05	~2.05	~2.317	NE	NE	~35	9
		10.1	NM[g]	NM	NM	NM	NM	NM	

[a] The molecular weights are in daltons

[b] The hyperfine constants $|A_i|$ are in units of $10^{-4}cm^{-1}$

[c] Azurin bleaches in alkaline and acid solutions, but the residual blue protein retains the same magnetic and optical properties as it gives way to the bleached protein

[d] About 60% of the 20,000 molecular weight of stellacyanin is protein, carbohydrate and hexoseamine each accounting for about 20% (1)

[e] $|A_{\parallel}^N| = 16$, $|A_{\perp}^N| = 12$ (6), where $\parallel$ means parallel to a Cu-N bond

[f] NE: Very small; quantitative estimate not available

[g] NM: Not yet measured

Table 3-4 (continued)

(b) Optical band at 10-12 kK

Protein	pH	λ(nm)[h]	$\bar{\nu}$[h]	w[i]	a[h]	$10^4 f$[j]	$10^{-3}[\theta]$[h]	$10^{39}R$[k]	Notes, refs.
azurin *Ps. aerug.*	neutral	~909	~11.0	~1.3	~0.35	~50	~-2.2	~-0.20	3
Stellacyanin	1-9	~909	~11.0	~1.3	~0.55	~75	~ 7.5	~ 1.5	3,10
	10-11.5	NM							
	12.8	NM							
Umecyanin	neutral	NM							
	9.5	NM							
	10.1	NM							

[h] λ is the wavelength and $\bar{\nu}$ is the wavenumber at the center of the band. a is the observed absorptivity and [θ], the observed ellipticity, at the center. There are contributions to a and [θ] from other bands, and a and [θ] are not necessarily maxima. $\bar{\nu}$ is in units of kK (i.e. $10^3 cm^{-1}$), a in $mM^{-1}cm^{-1}$, and [θ] in deg cm^2 $(decimol)^{-1}$

[i] w is the rms width of the optical band in kK. The bands are taken to be Gaussian, in which case the amplitude has dropped to $e^{-1/2}$ of the peak value at ± w away from $\bar{\nu}$

[j] f is the oscillator strength, given by $10^4 f = 0.108\ w\,a$ for a Gaussian band

[k] R is the rotational strength, given by $10^{39}R = 1.736\ (10^{-3}[\theta])(w/\bar{\nu})$ for a Gaussian band

Table 3-4 (continued)

(c) Optical band at 12-14 kK

Protein	pH	λ (nm)	$\bar{\nu}$	w	a	$10^4 f$	$10^{-3}[\theta]$	$10^{39} R$	Notes , refs.
azurin *Ps. aerug.*	neutral	764	13.1	~1.0	~0.95	~50	- 9.0	~-1.3	3
Stellacyanin	1-9	~790	~12.7	~1.1	~0.5	~50	~-20	~-2.9	3,10
	10-11.5	NM							
	12.8	NM							
Umecyanin	neutral	787	12.7	1.1	0.42	50	-11.4	-1.72	11
	9.5	NM							
	10.1	778	12.85	1.59	NM	-	-10.4	-2.24	11

Table 3-4 (continued)

(d) Optical band at 15-18 kK

Protein	pH	λ (nm)	$\bar{\nu}$	w	a	$10^4 f$	$10^{-3}[\theta]$	$10^{39}R$	Notes, Refs.
azurin *Ps. aerug.*	neutral	6.25	16.0	1.3	3.50	480	8.8	1.25	2,3,12
Stellacyanin	1-9	605	16.5	1.1	4.08	440	10.0	1.15	3,10
	10-11.5	NM							
	12.8	NM							
Umecyanin	neutral	606	16.5	1.06	3.36	385	3.30	0.370	11
	9.5	NM							
	10.1	581	17.2	1.41	NM	-	6.50	0.932	[ℓ],11

[ℓ]It is likely that the band at 581 nm for umecyanin at pH 10.1 is the superposition of two bands

(e) Optical band at 18-20 kK

Protein	pH	λ (nm)	$\bar{\nu}$	w	a	$10^4 f$	$10^{-3}[\theta]$	$10^{39}R$	Notes, Refs.
azurin *Ps. aerug.*	neutral	529	18.9	1.1	~0.60	~25	3.1	0.30	3,12
Stellacyanin	1-9	521	19.2	1.1	~0.4	~50	2.5	0.25	3,10
	10-11.5	NM							
	12.8	NM							
Umecyanin	neutral	524	19.1	1.13	0.46	56	1.75	0.181	11
	9.5	NM							
	10.1	NM							

Table 3-4 (continued)

(f) Optical band at 20-24 kK

Protein	pH	λ (nm)	$\bar{\nu}$	w	a	$10^4 f$	$10^3[\theta]$	$10^{39}R$	Notes, refs.
azurin *Ps. aerug.*	neutral	467	21.4	1.2	0.33	45	- 3.3	-0.35	2,3,12
Stellacyanin	1-9	448	22.3	1.3	0.80	110	-20	-2.0	3,10[m]
	10-11.5	NM							
	12.8	NM							
Umecyanin	neutral	451	22.15	1.24	0.37	49	- 8.60	-0.838	11
	9.5	NM							
	10.1	441	22.7	1.13	NM	-	- 7.55	-0.656	11

[m] The absorption is measured as the difference between the oxidized and reduced forms

References: (1) Ambler (1968); (2) Brill et al. (1968); (3) Hill (1972); (4) Peisach et al. (1967); (5) Malmström et al. (1970); (6) Rist et al. (1970); (7) Paul and Stigbrand (1970); (8) Stigbrand (1971); (9) Stigbrand et al. (1971); (10) Falk and Reinhammar (1972); (11) Stigbrand and Sjöholm (1972); (12) Tang et al. (1968)

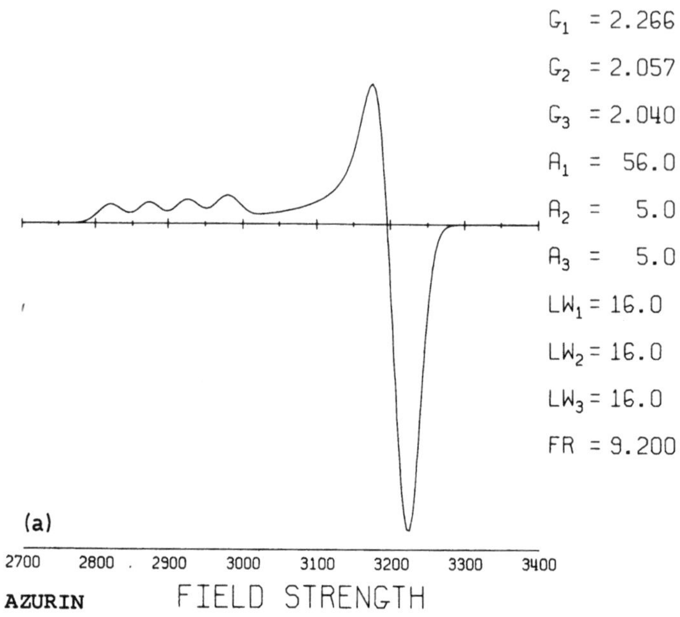

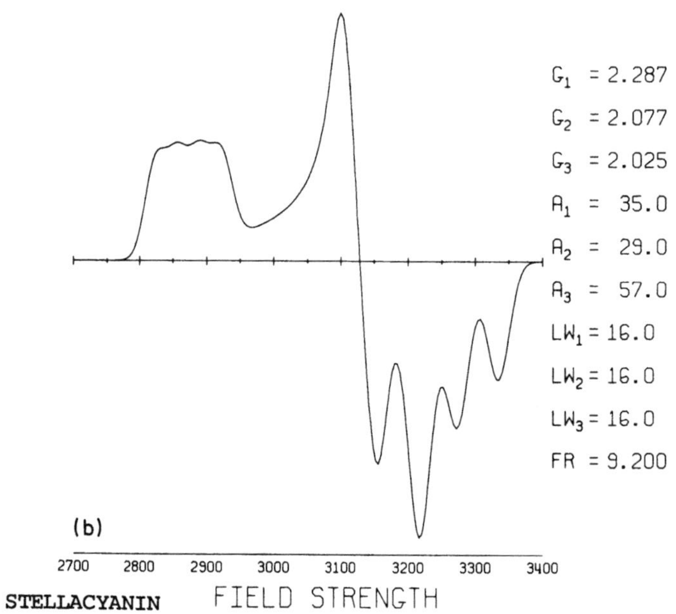

Fig. 3-5a-d. EPR, absorptivity, and circular dichroism spectra computed from the experimentally determined magnetic and optical parameters of azurin and stellacyanin; the G_i are g-values, A_i are A-values in units of 10^{-4} cm^{-1}, LW_i are linewidths in gauss, and FR is the microwave frequency in units of 10^9 Hz; the units of absorptivity and ellipticity are mM^{-1} cm^{-1} and deg cm^2 $(decimol)^{-1}$ respectively

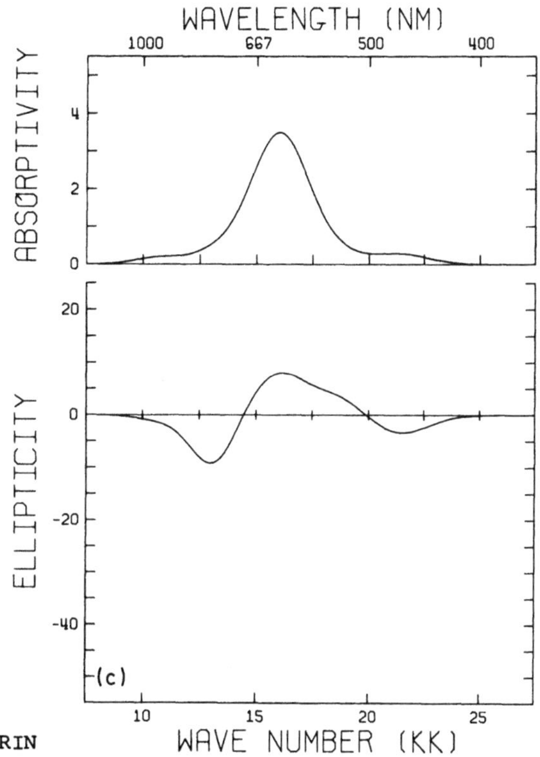

AZURIN

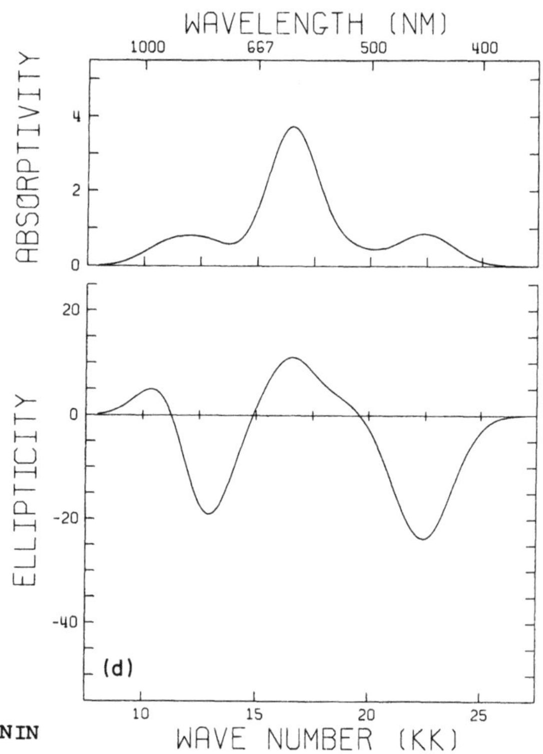

Fig. 3-5 c and d STELLACYANIN

Data from blue proteins containing several coppers is not included in Table 3-4. For example, there is fungal laccase which has one cupric site with $|A_z| = 90 \times 10^{-4}cm^{-1}$, a second cupric site with $|A_z| = 194 \times 10^{-4}cm^{-1}$, and also two coppers that are not detectable by EPR. In discussing EPR-detectable cupric ions, it is convenient to classify them into two kinds: Type 1 with $|A_z|$ less than $100 \times 10^{-4}cm^{-1}$ and Type 2 with $|A_z|$ greater than $140 \times 10^{-4}cm^{-1}$ (Vänngård, 1967). Thus far, Type 1 sites have been found only in proteins, and only in those proteins which have an intense absorption band at ~ 16 kK, namely the blue proteins. Type 2 sites are found for all other cupric complexes, including all protein sites that are non-"blue". There are very few exceptions to this generalization. Cytochrome oxidase contains two irons (in hemes) and two coppers. About half of the copper is EPR-detectable (see, however, Aasa and Vänngard, 1975) with parameters $g_{\perp} \approx 2.03$, $g_{||} = 2.17$, and $|A_{||}| \leq 30 \times 10^{-4}cm^{-1}$ (Beinert et al., 1962). It is difficult to comment upon the optical properties of the copper sites in cytochrome oxidase because of the presence of the heme groups which absorb intensely in the same regions as copper. A few small cupric complexes with nearly tetrahedral coordination have $|A_z|$ values which lie close to the upper end of the Type 1 range.

In view of the association of the intense blue color with Type 1 paramagnetic sites, the question arises as to how closely the optical and magnetic phenomena are related. Does, in fact, the blue color require the presence of the cupric ion? The cuprous ion does not exhibit EPR absorption. The integrated intensity of the EPR absorption spectrum from samples of equal volume is proportional to the concentration of cupric sites. Provided then that a decolorization process is available that is slow enough for measurements of optical and EPR spectra to be made on samples at the same well-defined extents of bleaching, the optical absorbance can be correlated with the concentration of cupric ion. An experiment of this kind was first done with ceruloplasmin, bleached with ascorbic acid, and it was found that the absorbance of the band at 16.4 kK (610 nm) is linearly proportional to the concentration of Cu(II) (Blumberg et al., 1963). However, the ceruloplasmin molecule is complicated, containing seven or eight coppers of which only four are EPR detectable, two of Type 1 and two (differing from each other) of Type 2 (Andréasson and Vänngård, 1970). Thus, there are several contributions to the EPR and optical spectra, and interpretation of the observation just given in terms of the behavior of the components requires more analysis than has yet been carried out. Fortunately the *Pseudomonas* azurins, which contain only one copper, of Type 1, per molecule, can be bleached very slowly by simply raising the pH above 8 (Brill et al., 1968; Maria, 1966). The decreases in optical and magnetic absorption are found to be linearly proportional. The EPR line shape and width remain constant, so that for these proteins the amplitude from samples of equal volume is proportional to the concentration of cupric sites. (It is not necessary to integrate the EPR absorption spectra.) It is clear then that the intense blue color is present only when the metal is in the cupric state. Do the factors which influence the magnetic parameters of the cupric ion in blue proteins also affect the optical properties? The best answer here is provided by the pH effects shown by stellacyanin (Peisach et al., 1967). From pH 2 to 8.3, the EPR spectrum and optical spectrum are constant. As the pH is raised above 8.3, both spectra change. In particular, the wavelength of maximum absorptivity of the most intense band shifts from 604 nm at neutral pH to 585 nm at pH 11 with virtually no change in intensity. At pH 11, the EPR spectrum shows $g_z = 2.312$ (as compared with 2.287 at neutral pH) and the low field copper hyperfine splitting is reduced to below $17 \times 10^{-4} cm^{-1}$ (Malmström et al., 1970). Umecyanin exhibits similar optical changes with pH (Stigbrand and Sjöholm, 1972), and there is some concomitant change in the EPR spectrum (Stigbrand et al., 1971) which has not yet been analyzed. In both stellacyanin and umecyanin, changes in optical activity occur along with the changes in optical absorption (and EPR). The azurin from *Pseudomonas aeruginosa* shows neither magnetic nor optical changes from pH 3 to pH 12, apart from the slow decolorization and loss of EPR signal which occurs in alkaline solution. The data at hand strongly suggest that alterations in the EPR spectrum are necessarily attended by changes in the optical properties, and therefore that there is a close relation between the structural features which give rise to the unusual absorption band at 16 kK and those which cause Type 1 hyperfine behavior. We will discuss first the contributions to the hyperfine splitting, looking for the molecular basis for Type 1 sites, and then turn to the molecular origins of the chromophoric behavior.

The interactions between the cupric electrons and the cupric nucleus can be written:

$$\mathcal{H}_{\text{Cu hyperfine}} = \mathcal{H}_{\text{Fermi}} + \mathcal{H}_{\text{orbital field-nuclear dipole}} + \mathcal{H}_{\text{electron dipole-nuclear dipole}} + \mathcal{H}_{\text{quadrupole}}.$$

$\mathcal{H}_{\text{orbital field-nuclear dipole}}$ is extremely unlikely to explain the anomalous A_z value. It is given by $P\,\underline{L}\cdot\underline{I}$ (Chap. 6-2). P will not vary greatly; $g_z - 2$, which is proportional to $\langle L_z \rangle$, is not unusual; the values of $\langle I_z \rangle$ are fixed. The major term of $\mathcal{H}_{\text{quadrupole}}$ does not have a first order effect on the low field resonance and this interaction cannot be offered as a cause of the small splitting. As discussed in Chapter 6-2, $\mathcal{H}_{\text{Fermi}}$ is an isotropic term which depends upon there being unpaired electron spin at the nucleus, a property possessed only by s orbitals. For normally bonded cupric ions, the amount of 4s mixing into the 3d orbitals is expected to be negligibly small. Rather one attributes the anomalously large isotropic hyperfine energy to core polarization wherein the spin "up" and spin "down" electrons of closed inner s-shells are unbalanced by interaction with the valence electrons (which have a net spin). The mechanism of core polarization is independent of the bonding of the cupric ion, and hence, apart from the effect of small differences in d-occupation attending delocalization and hybridization, the contribution of core polarization to the hyperfine energy will be constant. This is found to be the case in small cupric complexes where it has the value

$$-\kappa P \alpha^2 = -0.43 \times 0.38(\pm 0.002) \times 0.78(\pm 0.05)$$

$$= -0.0127(\pm 0.0014)\,cm^{-1}.$$

It is a curious feature of the Type 1 sites of the blue proteins that the $|A_z|$ values are depressed below the usual range by approximately 0.0110 to 0.0140 cm^{-1}. The ready explanation, namely $\kappa = 0$ for the blue proteins, is almost certainly wrong for the reason just given — κ is not expected to vary significantly from complex to complex. Rather an additional interaction or interactions have become significant in the blue proteins. Two likely possibilities are:

1. The admixture of a very small amount of 4s character to the hole orbital;

2. The admixture of a larger amount of 4p character.

The former mechanism is extremely effective. From the hyperfine structure of the $3d^{10}4s$ ^2S term of Cu (Tolansky and Forester, 1938), $A_F^{4s} \cong 0.2$ cm^{-1}. The entire $|A_z|$ displacement of the blue protein would then be accounted for by a 6% admixture of 4s. Use of much greater 4s character to explain the behavior of a 3d metal ion has been made by Adams et al. (1968) to explain a doubling of ~ 0.25 cm^{-1} in the emission spectrum of ScO [Sc(II) would be expected to be $(3p)^6 3d^1$]. With regard to the second mechanism above, Bates et al. (1962) have shown that p_z admixture to the d_{xy} orbital appropriate to the ground state of a tetrahedral site is effective in stabilizing this state and in reducing the copper hyperfine splitting. The interaction involved here belongs to $\mathcal{H}_{\text{electron dipole-nuclear dipole}}$, and the reduction arises because the magnetic field at the origin from spin in a p_z orbital is of opposite sign to that from spin in a d_{xy}, or $d_{x^2-y^2}$ for that matter. Brill and Bryce (1968) have used both mechanisms for reducing $|A_z|$ in their treatment of the cupric ion in blue proteins, 4s hybridization implicity through reduction in κ, and 4p hybridization explicitly. This kind of tetrahedrally distorted model for the cupric site in blue proteins, wherein the loss of a center of symmetry reduces hyperfine splitting and enhances optical absorption, had been suggested earlier (Brill et al., 1964).

The absorption spectra of the blue proteins from 10 kK to 25 kK show a very intense band at ~ 16 kK and two moderately intense bands at ~ 13 kK and ~ 22 kK. The latter two bands are more poorly resolved in the azurins than in stellacyanin (Fig. 3-5). In stellacyanin, the lowest energy band is broad and unsymmetrical which might suggest that it is the sum of bands, but this is not the case for the next two bands. Quantum mechanical models were developed which sought to match this data with single transitions assigned to the wing bands and transitions to two nearly degenerate states for the main band (Blumberg, 1966; Brill and Bryce, 1968). The circular dichroism studies of azurin and fungal laccase by Tang et al. (1968) provided the first clear evidence that there is a band of relatively weak oscillator and rotational strengths under the high energy limb of the main band, the center of the new band being at ~ 19 kK. This band has now also been found in the CD spectra of umecyanin (Stigbrand and Sjohölm, 1972), stellacyanin, ceruloplasmin, and tree laccase (Falk and Reinhammar, 1972), and the concomitant distortions of the low wavelength part of the main absorption band noted. Furthermore, in stellacyanin (Fig. 3-5) and tree laccase, the CD spectra reveal that the lowest energy absorption band (as hinted by its shape) includes two transitions, at ~ 11 kK and ~ 13 kK. Neither ceruloplasmin nor fungal laccase show two resolvable CD bands below 15 kK, but the CD band at ~ 13.5 kK is broad in both cases and accompanied by absorption which extends in energy below 10 kK. Additional CD measurements would be helpful to have for ceruloplasmin and for fungal laccase, and even more so for the simpler proteins umecyanin and azurin that have been studied with CD only down to 13 kK and 11 kK (Hill, 1972) respectively. Stellacyanin, umecyanin, and azurin all have a positive CD band at 33 kK, very much the same as cupric peptides in which metal-amide nitrogen bonding occurs (first noted by Martin, 1973). Because ceruloplasmin and the laccases have more than one kind of cupric site, their optical absorption and activity spectra are composite; for these proteins, straight-forward assignment of a set of bands to a particular site is not possible, even if the resolution of the spectra into bands were certain. For this reason, we will summarize now the optical data for those blue proteins containing only one cupric ion as shown in Table 3-4.

The salient optical features of Table 3-4 are:

1. The absorption bands of the three proteins are readily grouped according to energy. Within each group the energies at the centers of the bands vary by less than 5%, and it is both convenient and realistic to call each group "a band".

2. Five bands are clearly discernible in the visible and near infrared regions of the spectra. These will be labeled by the integral number of kiloKaysers nearest to the center of each group of bands: 11, 13, 16, 19, and 22 kK.

3. The peak absorptivity of each of these five bands greatly exceeds that of all the d-d transitions discussed earlier in this chapter.

4. The oscillator strength of the band at 16 kK is much greater than that of the other bands, and is about the same for all three proteins.

5. The strongest optical activity is shown by the band at 13 kK, the weakest by the band at 19 kK.

6. The sign of the rotational strength is either positive or negative only for the band at 11 kK.

7. The widths of the bands at 16 and 22 kK average 0.16 kK greater than the other bands.

In the absence of organic prosthetic groups, the energies of these bands suggest that all of the excited states include significant d-character. The large intensities suggest involvement of charge transfer with ligands disposed in low symmetry. The EPR parameters imply that the ground state is also of low symmetry, and the ENDOR data from stellacyanin (Rist et al., 1970), now to be discussed, require ground state delocalization.

None of the blue proteins exhibits resolved ligand hyperfine splitting in EPR spectra. However, ENDOR spectra from stellacyanin show that the cupric ion is coordinated to at least one nitrogen, and probably more than one, with $A_{\parallel}^{N} \approx 0.0016\ cm^{-1}$ and $A_{\perp}^{N} \approx 0.0012\ cm^{-1}$. It can be demonstrated that, if the coordinate sigma link involves an sp^2 trigonal hybrid on the nitrogen, $A_{\parallel}^{N}/A_{\perp}^{N} \approx 1.2$, and if an sp^3 tetrahedral hybrid, $A_{\parallel}^{N}/A_{\perp}^{N} \approx 1.3$. The accuracy of the A values from this pioneering ENDOR experiment is such that the observed ratio $A_{\parallel}^{N}/A_{\perp}^{N} = 1.33$ probably has an uncertainty of 5 or 10%. If the hybridization were tetrahedral, $A_{F}^{N} = 1.10\ A_{\perp}^{N} = 0.00132\ cm^{-1}$ and $(\alpha')^2 = 0.40$ for four similar ligands, an extreme delocalization; if trigonal, $A_{F}^{N} = 1.07\ A_{\perp}^{N} = 0.00128\ cm^{-1}$, and $(\alpha')^2 = 0.29$ for four such ligands, a moderate covalency. In either case, the ground state of the cupric ion-protein complex in stellacyanin is one in which a significant fraction of the unpaired spin is on the ligands. Proton magnetic resonance studies of plastocyanins suggest that the imidazole groups of two histidine residues are bound to the copper (Markley et al., 1975). In particular, peaks of the PMR spectra assigned to two histidine side chains disappear when the plastocyanins are oxidized.

Several kinds of information point to sulfur, in particular cysteinyl sulfur, as one of the copper ligands. Apostellacyanin has one titratable sulfhydryl group (McMillan et al., 1974; Morpurgo et al., 1972). The resonance Raman spectra from azurin, plastocyanin and ceruloplasmin all have a weak band at about 270 cm^{-1} which is likely to be associated with a copper-sulfur stretching mode. (Sulfur is about double the mass of nitrogen and oxygen which, as potential ligand atoms, could be responsible for the several Raman bands between 350 and 470 cm^{-1}; Miskowski et al., 1975.) X-ray photoelectron spectroscopy in the 2p binding energy region shows that one sulfur peak is shifted in both copper and cobalt plastocyanin from the sulfur band in the apoenzyme (Solomon et al., 1975). Much earlier suggestions of the proximity of cysteinyl sulfur to the copper were made on the basis that the reduction of Cu(II) at high pH was likely to be effected by oxidation of S^- (Brill et al., 1968; Maria, 1966).

Taken together, the magnetic, optical, and sequencing data from blue proteins suggest that the cupric ion is bound to three nitrogens (two imidazole, one peptide) and one cysteine sulfur in a complex of low symmetry.

In models for the cupric complex in blue proteins, delocalization of the ground and excited states need not be explicitly expressed in molecular orbitals which include ligand atomic orbitals. Rather little is lost by treating the ligand parts of the molecular orbitals as atomic orbitals of the central (copper) ion which have the same geometrical properties. This is the well-known "united atom" description of molecules. The cupric sites in blue proteins need not, and are likely not to belong to one of the point groups. However, the point group D_2, depicted in Figure 6-3, has a symmetry low enough to be consistent with most of the optical and magnetic properties of these special copper complexes. When s, p, and d orbitals are considered, the geometry of D_2 is such that eigenstates of the Hamiltonian can be mixtures

Fig. 3-6. Molecular orbital formation in D_2

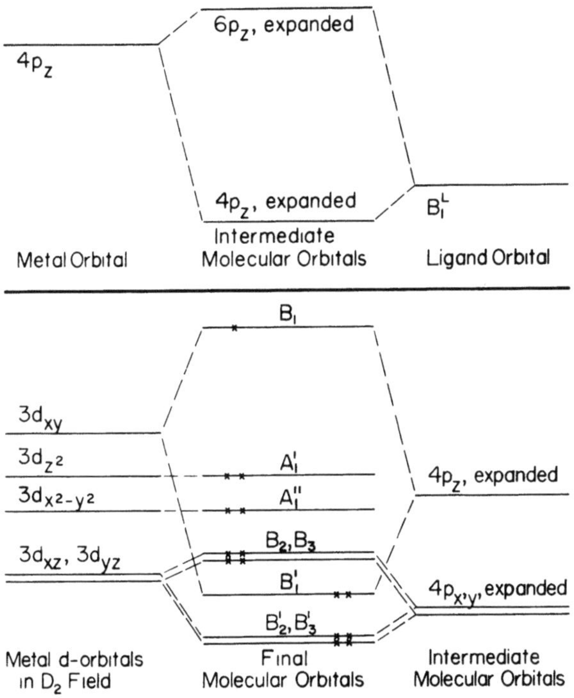

of d_{ij} and p_k, d_{z^2} and s, and $d_{x^2-y^2}$ and s. Thus, for example, the linear combination

$$|B_1\rangle = \alpha|3d_{xy}\rangle + \sqrt{1-\alpha^2}\;|4p_z\rangle$$

is a proper orbital for D_2; "B_1" is the conventional designation of orbitals of this angular behavior when D_2 point group symmetry applies. Normally, because of the high energy of the 4p orbitals relative to the 3d, the mixing of the two is very small in cupric systems. However, when ligand electron donor energy levels are high (comparable with p levels) and the ligands are spatially disposed such that they can interact with metal p orbitals, the resulting molecular orbitals have approximately the symmetry of p atomic orbitals, can be viewed as p orbitals with expanded radial functions, have lower and higher energies than the original levels, and the lower energy ones then mix more strongly with d orbitals. Figure 3-6 schematizes this sequence of interactions as it can occur for the cupric ion in D_2 symmetry (for simplicity the mixing of 4s with d_{z^2} and $d_{x^2-y^2}$ has been neglected). It is convenient to consider the energy levels of the "hole", shown in Figure 3-7, a description equivalent to that of the 15 electron system of Figure 3-6.

It has been shown (Brill and Bryce, 1968) that the assignments of (a) the intensely absorbing, relatively broad band at 16 kK to transitions from the hybrid $3d_{xy}$, $4p_z$ ground state B_1 hole orbital of Figure 3-7 to the closely energetic hybrid B_2, B_3 hole orbitals $3d_{xz}$, $4p_y$ and $3d_{yz}$, $4p_x$, and (b) the strongly

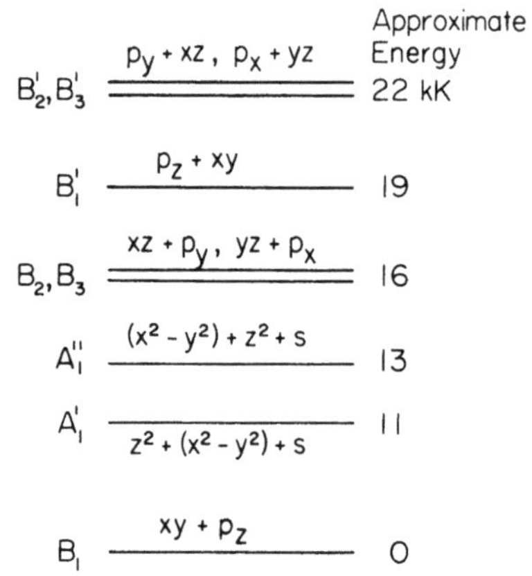

Fig. 3-7. Energy levels of the cupric ion hole in a site of D_2 symmetry

rotating band at 13 kK to a transition from the ground state to the A_1'' hole orbital primarily $3d_{x^2-y^2}$ in character, are consistent with the measured optical and magnetic parameters from azurins. Assignment (a) generates intense absorption through three factors: (1) the electric dipole operator r_i connects d_{ij} with p_j and d_{ik} with p_k; (2) the electric dipole matrix element between B_1 and B_2 (or B_3) then has two terms (since the oscillator strength goes as the square of the matrix element, the increase in intensity goes as the square of the number of terms of similar magnitude); (3) B_2 and B_3 are unresolved bands (the oscillator strengths of the transitions to these states are summed). Assignment (b) generates strong rotation by virtue of the large magnetic dipole matrix element between A_1'' and B_1. These assignments remain in agreement with recent more complete data from azurins and are also consistent with data from other blue proteins. At the time of the analysis of Brill and Bryce (loc. cit.), the bands at 11 and 19 kK were unresolved, and the expected transition to a $3d_{z^2}$ (A_1') hole was thought to be the band observed at 22 kK. From feature (6) (p. 59) one can infer that the band at 11 kK arises from a transition between states for which the magnetic (orbital) dipole moment vanishes at zero order. From the representation with respect to the d manifold, given in Table 6-1, one sees that the components of L are zero between $d_{x^2-y^2}$ or d_{xy} and d_{z^2}, and that all the other off-diagonal matrix elements are non-zero. It appears, therefore, that the band at 11 kK arises from a transition from the B_1 hole ground state (mainly d_{xy}) to the A_1' hole state (mainly d_{z^2}). The transitions to excited states which are primarily 3d in character are then accounted for by the bands at 11, 13 and 16 kK. The bands at 19 and 22 kK are associated with transitions from the ground B_1 to levels primarily 4p in character.

Quantitative aspects of the atomic hybrid model for the cupric sites in blue proteins will be discussed in the following section. The reader is referred to Blumberg (1966) for a description of the crystal field treatment of this kind of site.

4. Magnetic and Optical Properties of Quantum Mechanical Models of the Cupric Ion[3]

It is convenient, at the start, to set up hybrid atomic orbital wavefunctions, the mixtures being appropriate to the point group D_2 except for the ground B_1 state which will be modified to include a small amount of 4s (for its hyperfine effect) character (not permitted for B_1 in D_2) and a smaller admixture of $3d_{x^2-y^2}$ to keep $|B_1\rangle^{mod}$ orthogonal to the A_1 states. The properties of a purely d manifold can then be obtained by setting certain of the hybridization parameters equal to 1 and others to 0, and the properties of a site of symmetry higher than D_2 can be found by making certain energies and hybridization parameters equal. According to the assignments of the preceding section, the hole orbitals (labeled as per D_2) for the cupric ion in blue proteins and the corresponding energies above the ground state, are in order of decreasing energy:

$$\left.\begin{aligned} |B_3'\rangle &= \sqrt{1-\beta_3^2}\,|3d_{yz}\rangle - \beta_3\,|4p_x\rangle \\ |B_2'\rangle &= \sqrt{1-\beta_2^2}\,|3d_{xz}\rangle - \beta_2\,|4p_y\rangle \end{aligned}\right\} \quad 22\text{ k}$$

$$|B_1'\rangle = \frac{\rho\sqrt{1+\sigma^2}}{\sqrt{1+\rho^2(1+\sigma^2)}}\,|3d_{xy}\rangle - \frac{1}{\sqrt{1+\rho^2(1+\sigma^2)}}\,|4p_z\rangle \qquad 19$$

$$\left.\begin{aligned} |B_3\rangle &= \beta_3\,|3d_{yz}\rangle + \sqrt{1-\beta_3^2}\,|4p_x\rangle \\ |B_2\rangle &= \beta_2\,|3d_{xz}\rangle + \sqrt{1-\beta_2^2}\,|4p_y\rangle \end{aligned}\right\} \quad 16$$

$$|A_1''\rangle = \frac{1}{\sqrt{(1+\delta^2)(1+\gamma^2)}}\,|3d_{x^2-y^2}\rangle - \frac{\delta}{\sqrt{1+\delta^2}}\,|3d_{z^2}\rangle + \frac{\gamma}{\sqrt{(1+\delta^2)(1+\gamma^2)}}\,|4s\rangle \qquad 13$$

$$|A_1'\rangle = \frac{\delta}{\sqrt{(1+\delta^2)(1+\gamma^2)}}\,|3d_{x^2-y^2}\rangle + \frac{1}{\sqrt{(1+\delta^2)}}\,|3d_{z^2}\rangle + \frac{\gamma\delta}{\sqrt{(1+\delta^2)(1+\gamma^2)}}\,|4s\rangle \qquad 11$$

$$|B_1\rangle^{mod} = \frac{1}{\sqrt{(1+\rho^2)(1+\sigma^2)}}\,|3d_{xy}\rangle + \frac{\rho}{\sqrt{1+\rho^2}}\,|4p_z\rangle + \frac{\sigma}{\sqrt{(1+\rho^2)(1+\sigma^2)}}\,|4s\rangle - \frac{\gamma\sigma}{\sqrt{(1+\rho^2)(1+\sigma^2)}}\,|3d_{x^2-y^2}\rangle \qquad 0$$

where the $|4s\rangle$ and $|4p\rangle$ orbitals are understood to have expanded radial functions. States B_1, A_1', A_1'', B_2, and B_3 are primarily d-states; B_1', B_2', and B_3', primarily p-states.

---

[3]The research of this section was supported by a grant from the National Science Foundation. The computer programs were written by Patricia H. Brill

We explore the magnetic properties of the d-manifold alone by taking the states B_1', B_2', and B_3' to be of such high energy (compared with the spin-orbit coupling constant) that they have no effect upon g- and A-values. The parameters of various models and the resulting EPR spectra are given in Table 3-5 and Figure 3-8. In these computations, the energy level differences between the ground state and the excited levels are as given above; the spin-orbit coupling constant (λ) and the isotropic contact term (κ) for $|3d\rangle$ are assigned the usual values of $-828\ cm^{-1}$ and 0.43 respectively; and the Fermi coefficient of the $|4s\rangle$ orbital is taken as $0.20\ cm^{-1}$. For the isolated cupric ion, the spin-orbit coupling constant (λ') for $|4p\rangle$ has the value $925\ cm^{-1}$, and this is used.

Case a is that of the pure d-manifold. Model b differs from a in that each d orbital is multiplied by 0.90 to show the effects of 19% delocalization into ligand orbitals which in themselves do not contribute to the magnetic parameters. Case c demonstrates that the introduction of (16%) p_z character into the ground state reduces the magnitude of the major copper hyperfine splitting (by over 40%), and has significant effects upon $A_\perp$ and the g-values as well. The connection of the $|3d_{x^2-y^2}\rangle$ part of state $|A_1''\rangle$ to the ground state by L_z gives rise to the major orbital component of $g_{||}$. The latter component and therefore $g_{||}$ are reduced in case d where the admixture of $|4s\rangle$ into $|A_1''\rangle$ when $\gamma \neq 0$ diminishes the $(3d_{x^2-y^2})$ term. In case d, the increase in $A_{||}$ follows from the decrease in $g_{||}$. Model e shows that the effect of hybridizing the states $|B_2\rangle$ and $|B_3\rangle$ is mainly upon $g_\perp$ and $A_\perp$. Introduction of 4s hole character into the ground state changes the hyperfine splittings markedly, case f. Combining the factors above, with the exception of the kind of delocalization in case b, one arrives at model g and an EPR spectrum much like that of azurin and umecyanin. Case h demonstrates that the latter kind of EPR spectrum can also be produced by larger admixtures of $4p_z$ and 4s into the ground state, with concomitant reduction of 4s in the A_1 states; it should be noted that the sign of $A_{||}$ for this model is positive. Whether it is model g or h which applies to azurin and umecyanin can be determined from the optical absorption spectra, in a manner to be discussed later. In the very rhombic EPR spectrum from stellacyanin, the largest hyperfine splitting is associated with the highest field resonance (smallest g-value). An axial approximation to this is case i. Just as for the azurin EPR spectrum, a second axial approximation with an A-value sign difference exists, but the rhombic spectrum derived from the latter set of parameters has the largest hyperfine splitting associated with the middle g-value and can be rejected on that basis. A proper rhombic model for stellacyanin is case j.

The inclusion of the states B_1', B_2' and B_3' in the calculation of the spin-Hamiltonian parameters does not significantly alter the conclusions about the magnetic properties of the models presented above. A discussion of optical properties will now be given, and it is necessary for this purpose to consider the entire set of eight states. It is, however, instructive at first to examine the oscillator strengths of the following simplified system:

Table 3-5. Magnetic properties of quantum mechanical models of the cupric ion

Case	Hybridization parameters[a]					g-values			A-values		
	β_2	β_3	γ	ρ	σ	$\perp$ or x	y	$\parallel$ or z	$\perp$ or x	y	$\parallel$ or z
a	1.0	1.0	0	0	0	2.106		2.512	-22.7		-161.1
b[b]	1.0	1.0	0	0	0	2.070		2.337	-22.9		-161.2
c	1.0	1.0	0	0.4	0	2.092		2.442	-41.6		- 94.8
d	1.0	1.0	0.6	0	0	2.106		2.337	-21.1		-208.1
e	0.84	0.84	0	0	0	2.075		2.512	-31.3		-165.8
f	1.0	1.0	0	0	0.2	2.102		2.492	55.1		- 78.0
g	0.84	0.84	0.6	0.4	0.2	2.060		2.313	13.9		- 58.2
h	0.84	0.84	0.3	0.6	0.25	2.046		2.326	14.1		56.9
i	0.84	0.84	0.6	0.6	0.14	2.048		2.272	-44.4		- 29.8
j[c]	0.92	0.76	0.6	0.6	0.14	2.065	2.034	2.272	-37.9	-49.5	- 30.3

[a] $\delta = 0$ in all cases

[b] Each d orbital multiplied by 0.90

[c] Δ_2 = 15.5 kK and Δ_3 = 16.5 kK

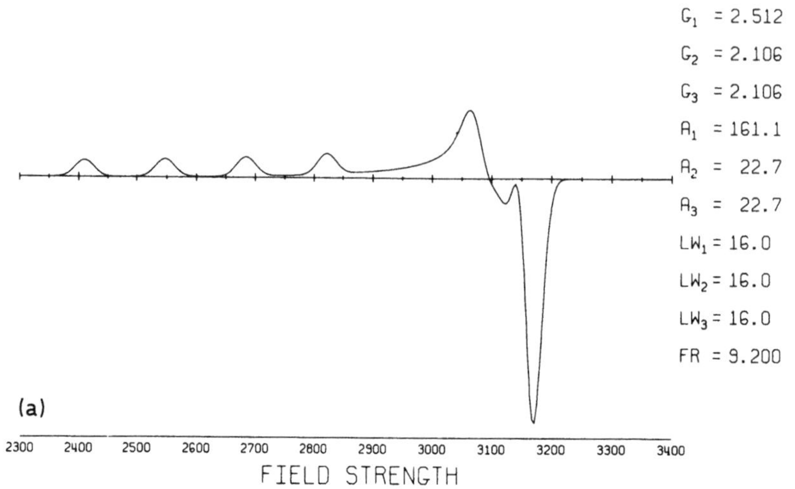

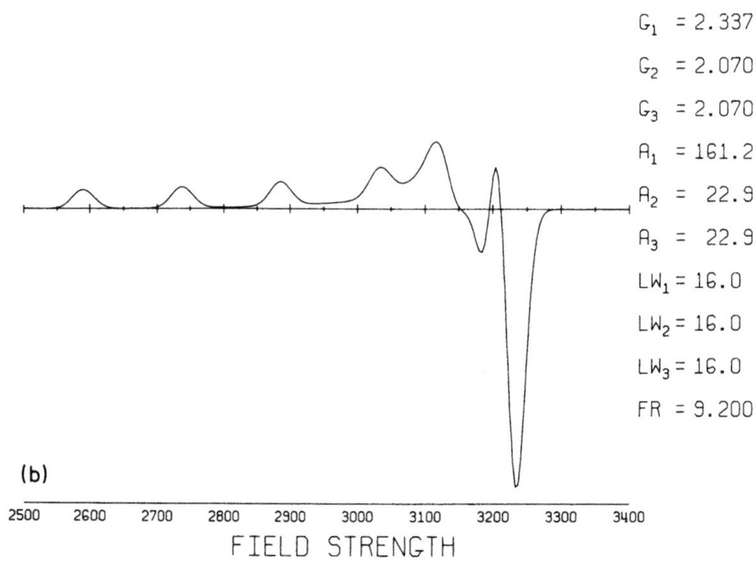

Fig. 3-8a-j. EPR spectra computed from the g-values and A-values of the ten models (a-j) of Table 3-5

$$\left.\begin{array}{l} |B_3'\rangle = \sqrt{1-\beta^2}\,|3d_{yz}\rangle - \beta\,|4p_x\rangle \\ |B_2'\rangle = \sqrt{1-\beta^2}\,|3d_{xz}\rangle - \beta\,|4p_y\rangle \end{array}\right\}\ 22\text{ kK}$$

$$\left.\begin{array}{l} |B_3\rangle = \beta\,|3d_{yz}\rangle + \sqrt{1-\beta^2}\,|4p_x\rangle \\ |B_2\rangle = \beta\,|3d_{xz}\rangle + \sqrt{1-\beta^2}\,|4p_y\rangle \end{array}\right\}\ 16$$

$$|B_1\rangle = \beta\,|3d_{xy}\rangle + \sqrt{1-\beta^2}\,|4p_z\rangle \qquad 0$$

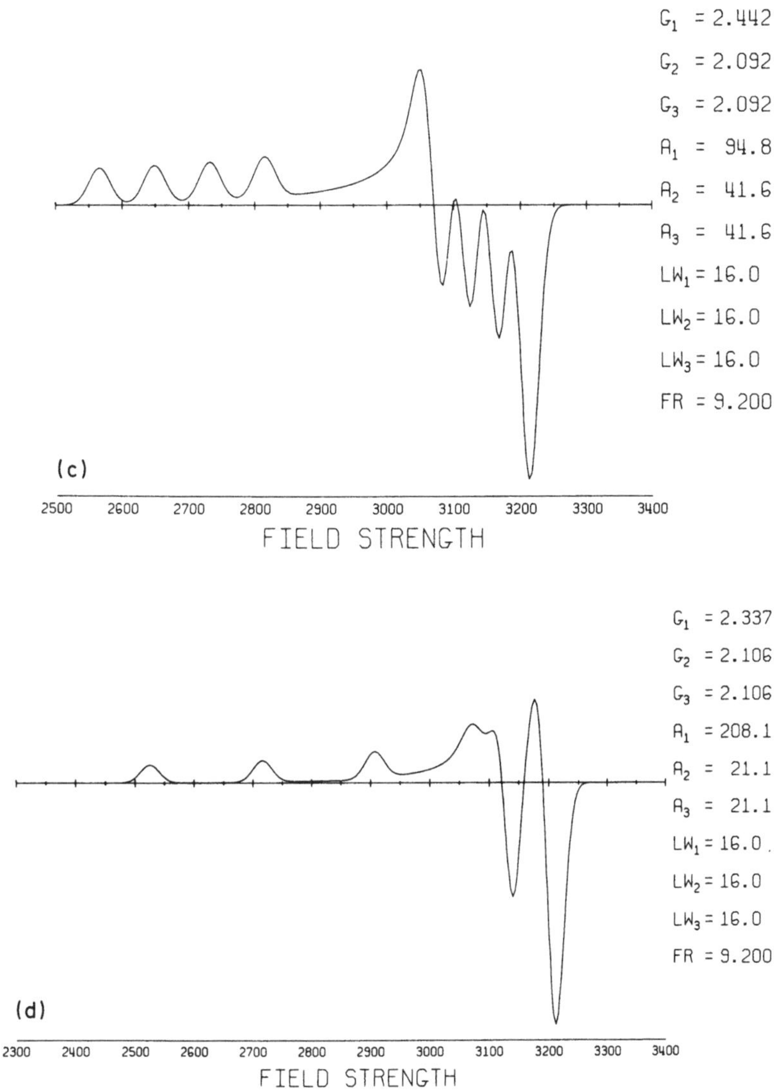

Fig. 3-8c and d (legend see opposite page)

It can readily be shown that

$$f_{16kK}/f_{22kK} = (16/22)\{4\beta^2(1-\beta^2)/[1-4\beta^2(1-\beta^2)]\},$$

a function graphed in Figure 3-9. When $\beta \approx 1$, the 22 kK levels are essentially ligand and f_{22kK}/f_{16kK} is large; this is the typical case where the presence of charge transfer bands is not accompanied by a marked increase in the intensity of d-d transitions. In the blue proteins, the opposite situation applies; for stellacyanin, umecyanin, and azurin the absorption of the 16 kK

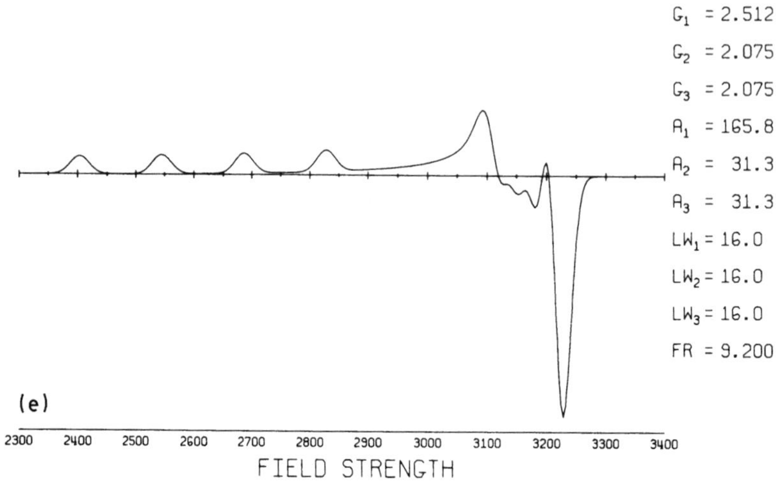

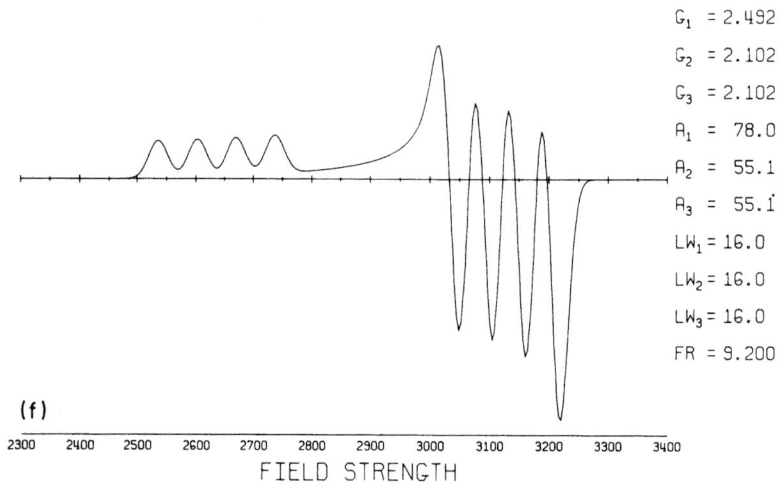

Fig. 3-8 e and f (legend see page 66)

band is stronger than the one at 22 kK by factors of about four, eight, and 13 respectively. Referring to Figure 3-9, one sees that β is in the region 0.80 to 0.85 for the blue proteins, close to the high end for stellacyanin and at the low end for azurin. In terms of the ground state hybridization parameters introduced earlier, the simplified optical system suggests that, since σ is small and $\beta \approx (1 + \rho^2)^{-1/2}$, ρ is in the range 0.6-0.75. This is found to be true for the complete system of states and rules out case g of Table 3-5 as a model for the EPR behavior of the cupric ion in azurin and umecyanin. The ground state orbital

$$\frac{1}{\sqrt{1 + \rho^2}}\left|3d_{xy}\right\rangle + \frac{\rho}{\sqrt{1 + \rho^2}}\left|4p_z\right\rangle$$

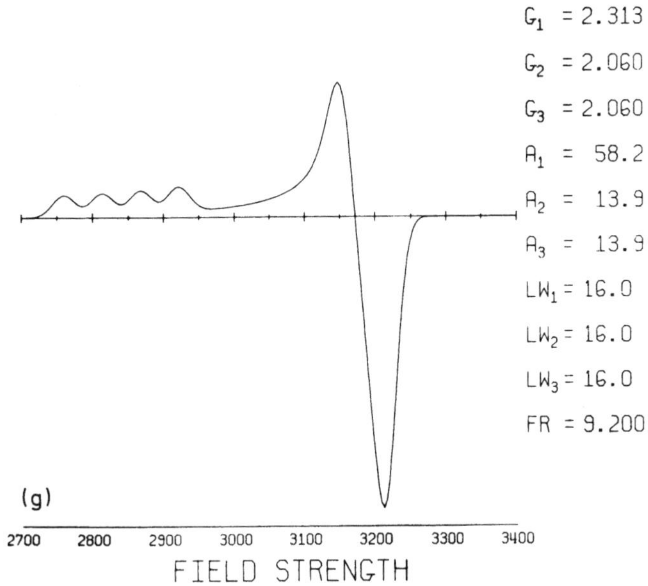

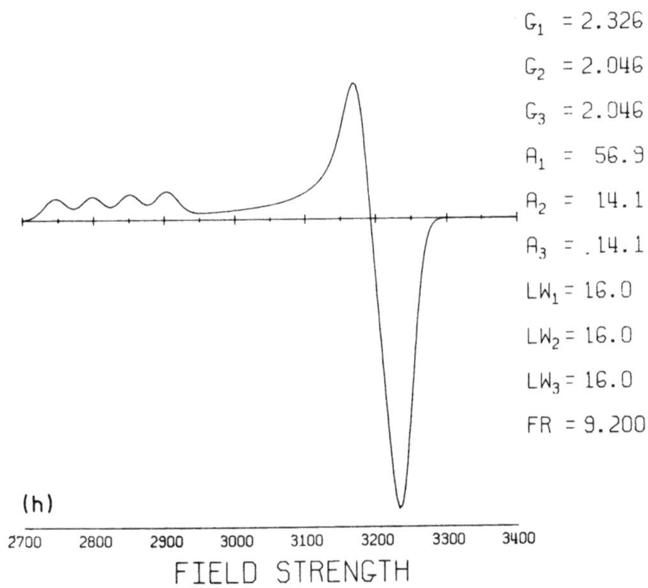

Fig. 3-8 g and h (legend see page 66)

has maxima in concentration at an angle η above and below the xy plane given by (Brill and Bryce, 1968)

$$\eta = 90^\circ - \cos^{-1}(\rho/\sqrt{5}) = \sin^{-1}(\rho/\sqrt{5}).$$

In Figure 3-10, the out-of-planarity angle η is shown as a function of the hybridization parameter ρ.

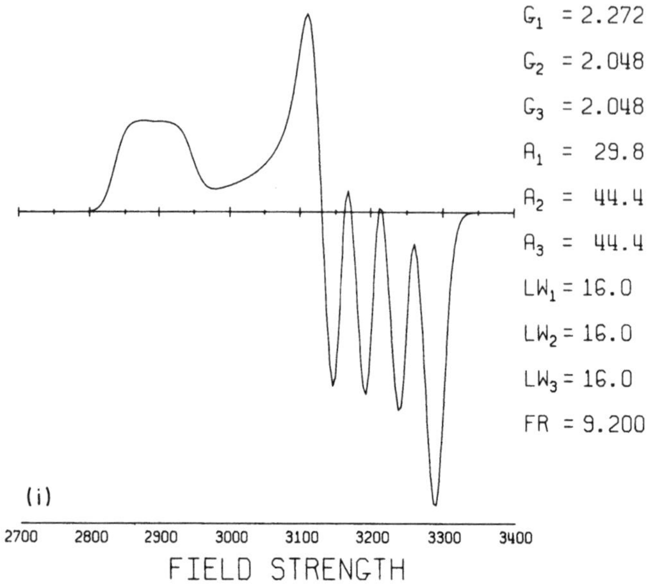

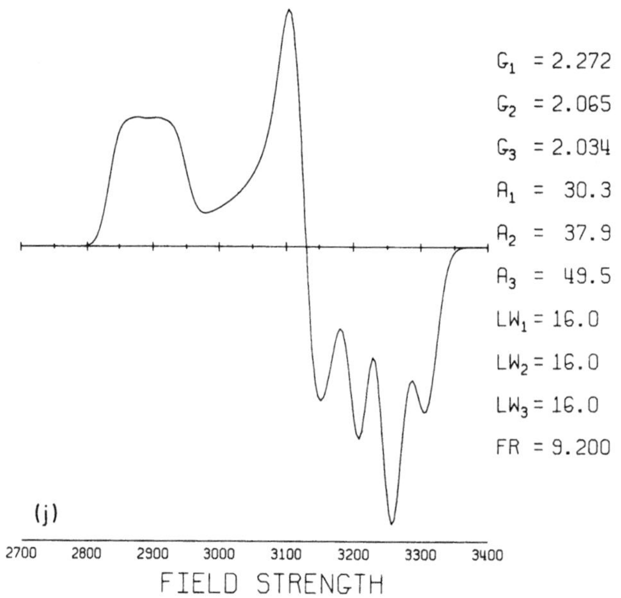

Fig. 3-8 i and j (legend see page 66)

The main features of the optical absorption spectra of the entire set of states are generated by axially symmetric models, such as the first four of Table 3-6, with corresponding absorptivity curves in Figure 3-11. In these calculations, the radial integrals $3d\langle r\rangle 4p$ and $4s\langle r\rangle 4p$ have been given the values 0.8 Å and -0.6 Å respectively which lead to oscillator and rotational strengths of

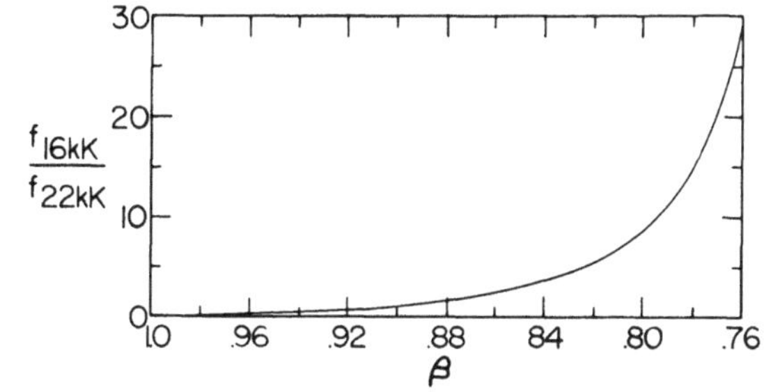

Fig. 3-9. Ratio of oscillator strengths for transitions between B levels

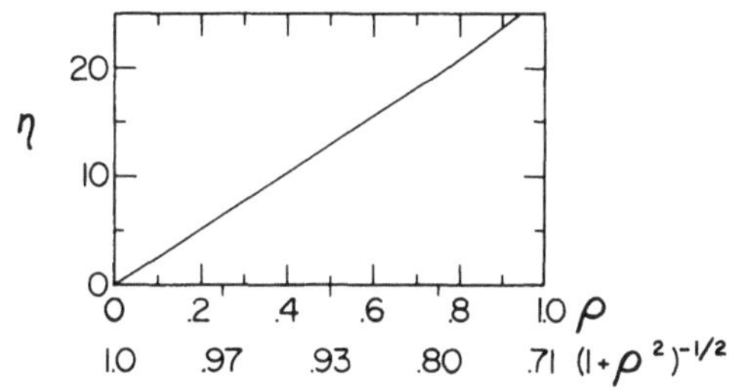

Fig. 3-10. Out-of-planarity angle η (in degrees) of B_1 orbital concentration as a function of 4p character

the correct magnitude.[4] In order for the rotational strengths to be of correct sign, the product of γ and $4s\langle r\rangle 4p$ must be positive. Since calculations of $4s\langle r\rangle 4p$ for atoms and ions show this integral to be consistently negative (a feature which can be explained in terms of the opposite signs of the major parts of 4s and 4p at large r), γ must be negative. (The signs of the hybridization parameters do not enter the calculation of the spin-Hamiltonian parameters and hence have all been taken positive in Table 3-6. γ negative implies that the A_1 orbitals which appear in these models are anti-bonding, a situation which can be viewed within the framework of Figure 3-7 when intermediate molecular orbitals of 4s symmetry are included. Apart from the fact that $(1 + \rho^2)^{-1/2} = 0.86$ rather than 0.84, case A belongs to the simplified system of the preceding paragraph. Model A exhibits no optical activity, but renders a realistic absorption spectrum. The effects of admixing 4s into A_1'', into the ground state, and then into both are shown by cases B, C, and D. Note the characteristic strong negative ellipticity which attends the introduction of a transition moment to state A_1'' (band at 13 kK); the large magnitude of $\langle A_1''|L_z|B_1\rangle^{mod}$ is responsible. The major contributions to the rotational strengths of the transitions to states B_2 and B_3 (and to B_2' and B_3' as well) are equal and opposite in axial models. When rhombicity is introduced $(\beta_2 \neq \beta_3)$, the magnitudes of the major contributions are no longer equal, and a net rotational strength appears; as can be seen in model E, this mechanism is capable of producing large

[4]These values for the radial integrals can be compared with those of Michel-Calendini and Kibler (1968) for Cu(II), $\langle 3d|r|4p\rangle = 0.31$ Å, and with those of Fliflet and Kelly (1973) for Zn(II), $\langle 3d|r|4p\rangle = 0.39$ Å and $\langle 4s|r|4p\rangle = -1.12$ Å

Table 3-6. Optical properties of quantum mechanical models of the cupric ion

Case	Hybridization parameters						Oscillator Strength × 10^4 of band at (in kK)					Rotational Strength × 10^{39} of band at (in kK)				
	β_2	β_3	γ	ρ	σ	δ	11	13	16	19	22	11	13	16	19	22
A	0.84	0.84	0	0.6	0	0	54	0	358	0	119	0	0	0	0	0
B	0.84	0.84	-0.6	0.6	0	0	54	12	358	0	119	0	-6.01	0	0	0
C	0.84	0.84	0	0.6	0.2	0	54	0	354	5	123	0	0	0	0	0
D	0.84	0.84	-0.6	0.6	0.14	0	54	12	356	3	121	0	-5.95	0.12	0.12	-0.12
E	0.92	0.76	-0.6	0.6	0.14	0	54	12	344	3	138	0	-5.95	2.88	0.12	-2.88
F	0.92	0.76	-0.6	0.6	0.14	-0.1	49	18	344	3	138	1.31	-7.25	2.88	0.12	-2.88
G	0.84	0.76	-0.3	0.6	0.25	+0.01	54	3	373	8	97	-0.15	-3.48	1.68	0.18	-1.68

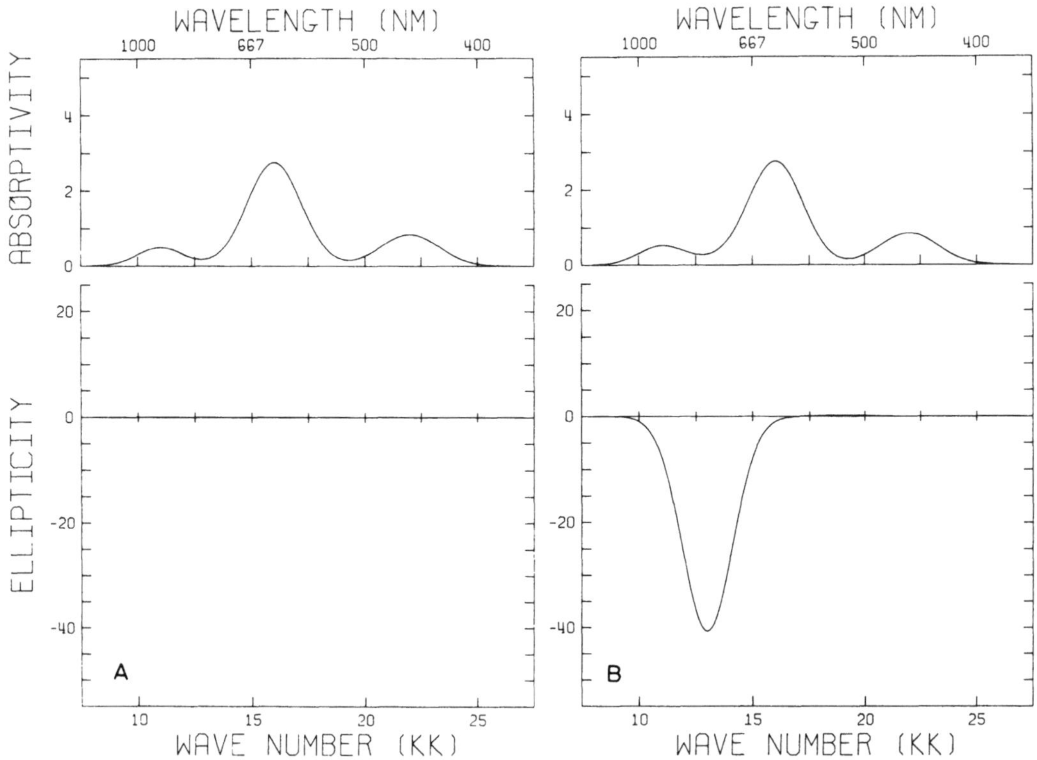

Fig. 3-11A-G. Absorptivity and circular dichroism spectra computed from the oscillator strengths and rotational strengths of the seven models (A-G) of Table 3-6

ellipticities. When δ as well as γ is nonzero, A_1' acquires some $3d_{x^2-y^2}$ character and with it a nonvanishing matrix element $\langle A_1'|L_z|B_1\rangle^{mod}$. The sign of the resulting rotational strength (band at 11 kK) depends upon the sign of δ; examples are given in cases F and G, the former a model for stellacyanin and the latter, for azurin.

The hybridization parameters of cases F and G of Table 3-6 are essentially the same as those of cases j and h of Table 3-5 which produce stellacyanin- and azurin-like EPR spectra respectively. (g- and A-values generated with the small δs of F and G differ negligibly from those with δ = 0.) Comparison with Figure 3-5 shows that the absorption and CD spectra of models F and G exhibit the main features of the spectra from stellacyanin and azurin, including the differences between the two proteins. Among the five rotational strengths which characterize each site, the model values are exact (11 kK) or within factors of 2(19 kK), 3(13 and 16 kK), and 4 1/2 (22 kK) of the protein values; three of the five oscillator strengths agree with the protein data within factors of 3. Clearly the geometrical features of the orbitals which determine the EPR spectra are also largely responsible for the characteristic optical properties of the

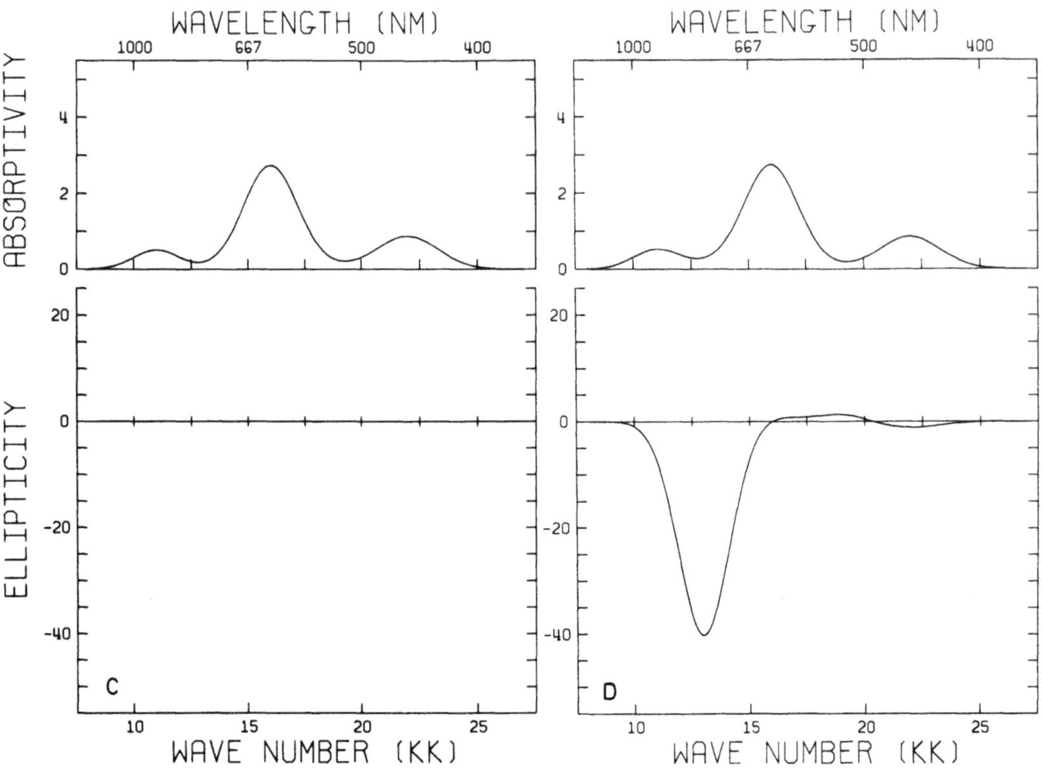

Fig. 3-11 C and D (legend see page 73)

cupric site in blue proteins. In the treatment above, only eight parameters (six hybridization and two radial matrix elements) have been adjusted to bring the calculated magnetic and optical properties into approximate agreement with the 16 experimentally determined parameters. Additional hybridization, and variation in other matrix elements (e.g. $\langle 1/r^3 \rangle$), would allow accurate fitting of the optical absorbance and CD spectra as well as the EPR spectra, but the critical features of the states of the cupric ion are already present in the simple models considered above. These features are:

1. The symmetry of the cupric sites is, to a first approximation, that of the point group D_2. This means that the usual planar geometry of cupric complexes is tetrahedrally distorted (significantly so), and there is also a substantial rhombic component.

2. Ligand electron donor orbitals are present with energies and spatial disposition such that they interact strongly with metal orbitals.

3. Extensive delocalization of metal electrons onto ligand atoms occurs, in the ground state and in the excited states; conversely, ligand electrons are present in substantial density on the copper.

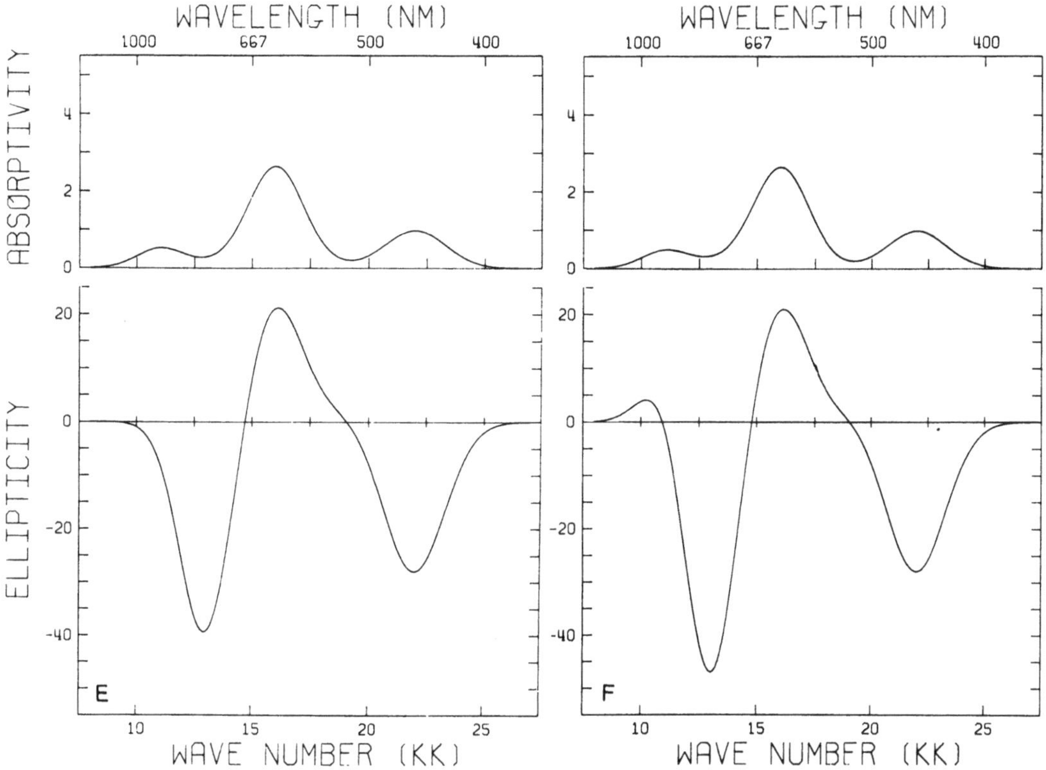

Fig. 3-11 E and F (legend see page 73)

This picture explains the low energies and high probabilities of transitions induced by near infrared and visible radiation. It also suggests the reason for the high reduction potential for the cupric-cuprous couple in blue proteins; the $3d^9$ states (e.g., $|B_1\rangle^{mod}$) are unusually close in energy and geometry to $3d^{10}$ states (e.g., $|B_1'\rangle$). Protein conformation has a role here in determining the redox potential, for displacement of the ligand atoms will affect the hybrid orbitals; dependence of spectral and oxidation-reduction properties upon pH would be expected on this basis. In addition, the delocalized hole density brings the reducing site volume closer to the protein surface. (The metal ion is known from proton relaxation data to be internally situated in plastocyanin, fungal laccase, and stellacyanin; Blumberg and Peisach, 1966; Malmström et al., 1968; Rist et al., 1970.)

Utilization of spectroscopic data from blue proteins for quantitative estimates of the bond angles awaits appearance of the X-ray-determined structures of these macromolecules. Certain correlations are suggested by the properties of the hybrid atomic orbital models; e.g. a measure of the tetrahedral distortion (out-of-planarity angle) may be the ratio of the oscillator

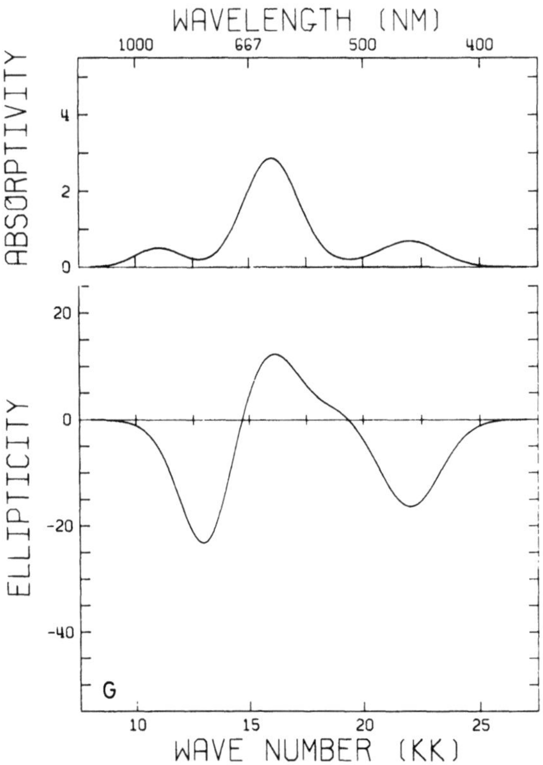

Fig. 3-11 G (legend see page 73)

strength of the band at 16 kK to the one at 22 kK. When such correlations have been established, it will be possible to interpret spectroscopic changes during reactions and under other unstable conditions in terms of changes in site geometry.

The "resting" Type 1 EPR spectrum appears to reflect a distribution in site geometry. There are obvious differences in the shapes and relative heights of the bands of the low field quartet of copper hyperfine lines between Figures 3-4 and 3-5 (azurin). In addition, the experimental spectrum (Fig. 3-4), upon careful examination is seen to have unequal peak separations (55, 56, and 48G) while the simulated spectrum (Fig. 3-5) has equal ones (53G). Realistic simulation of the experimental spectrum can be achieved by summing spectra generated with spin-Hamiltonian parameters calculated over a Gaussian distribution in tetrahedral angles. The result of a model computation is shown in Figure 3-12 where β, γ, and σ are given the values of case h of Table 3-5 (Fig. 3-8 h), and the distribution $N(\rho) = \exp[-(\rho - \rho_o)^2/2(\Delta\rho)^2]$ is taken with $\rho_o = 0.6$ (as in case h) and $\Delta\rho = 0.026$. This value of $\Delta\rho$, which among a set of $\Delta\rho$ provides spectral features closest to those of Figure 3-4, corresponds to a root mean square distribution in out-of-planarity angle η (Fig. 3-10) of 0.012 radian (0.7°).

Fig. 3-12. EPR spectrum computed on the basis of a Gaussian distribution in tetrahedral distortions of rms deviation 0.7° about model h of Table 3-5

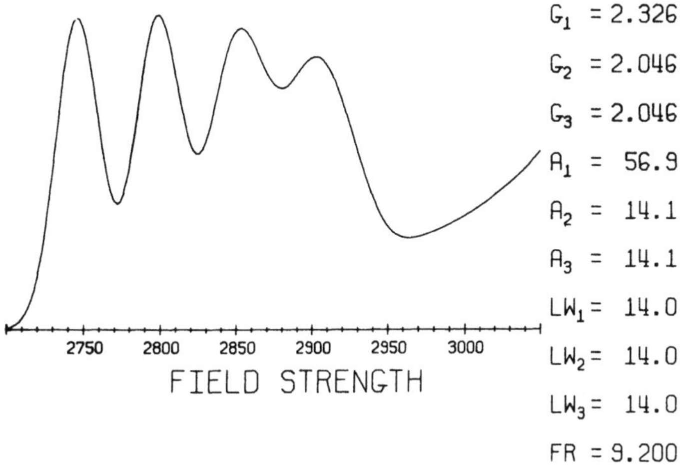

5. "Nonblue" Coordination in Copper Proteins

From the ground state of the cuprous ion d-d transitions cannot occur and, unless there is a low-lying ligand electron-acceptor state which would result in a charge transfer transition, a cuprous complex will be colorless. Furthermore, the ground state of the cuprous ion is diamagnetic and does not show EPR absorption. Clearly then, cuprous states of the copper-binding sites in proteins will not exhibit the special spectroscopic properties of the blue proteins. However, many copper-containing proteins are not blue when the metal is in either valence state, and in most of the blue proteins which have two or more copper ions per molecule, not all of the sites have the intense blue chromophore and Type 1 EPR behavior even when cupric.

At this time, two kinds of "nonblue" copper coordination are recognized in proteins, "EPR-detectable" and "EPR-nondetectable". Apart from the special case of cytochrome oxidase, there are many examples of nonblue cupric sites of the former kind — e.g. in the enzymes superoxide dismutase, galactose oxidase, and dopamine-β-hydroxylase — all of which have EPR spectra of Type 2 with $|A_{\|}|$ values in the range 0.015 to 0.020 cm^{-1}. The magnetic properties of this class of sites are not unusual, and, apart from the visible absorption and ellipticity of superoxide dismutase being somewhat greater than that of cupric peptides, the optical properties have no special features either. It is interesting to note that, upon cooling to 77 K, the peak absorptivity (at ~ 680 nm) of bovine superoxide dismutase increases five times over that at room temperature (Weser et al., 1971), bringing it within a factor of about three of the peak absorptivities of the 16 kK (~ 620 nm) band of the blue proteins.

The question arises as to whether or not the valence of the copper ion in EPR detectable sites in enzymes cycles during the catalysis of oxidation-reduction reactions. There is some evidence that such valence changes may take place, for the amplitude of the Cu(II) EPR signal from dopamine β-hydroxylase is significantly diminished in the presence of reducing substrate (Blumberg et al., 1965a; Friedman and Kaufman, 1966), and both the Cu(II) EPR signal and visible absorption of superoxide dismutase show decreases in the presence of superoxide radical (O_2^-) (Rotilio et al., 1972). However, EPR changes of this kind have not been found for the cupric sites in the amine oxidases or in galactose oxidase (Blumberg et al., 1965b).

When chemical analysis yields a greater number of copper atoms than integration of EPR absorption, assignment of the difference to cuprous ions is the simplest but not the only explanation. If in addition the magnetic susceptibility is measured and no paramagnetic moment is found, the EPR nondetectable copper is likely to be univalent, but it is also possible that cupric ions can be involved in strong antiferromagnetic exchange (Kokoszka and Gordon, 1969). In the isolated pairs of coupled metal ions in cupric acetate and other Cu(II) carboxylate dimers, for example, the upper triplet state is 300 cm^{-1} (equivalent to ~ 440 K) or higher above the ground singlet, and these compounds would be expected to be essentially diamagnetic over a large range of temperatures. It has been found that super exchange via the carboxylate bridges (rather than metal-metal bonding) is responsible for the spin coupling in the Cu(II) carboxylate dimers (Goodgame et al., 1969). The EPR spectra from exchange-coupled and non-interacting cupric ions will differ. Dipolar interaction without exchange can also occur between cupric ions and results in EPR line broadening and extra resonances at low field. Observation of the latter effects in (frozen) solutions of peptide complexes has been interpreted as reflecting dimerization (Boas et al., 1969). The copper-containing proteins tyrosinase and hemocyanin, which in the native state exhibit very little or no EPR absorption, are EPR-active in the presence of NO (Schoot Uiterkamp and Mason, 1973). One of the two major Cu(II) EPR signals from NO-treated tyrosinase and hemocyanin shows the typical properties of pairs of magnetic dipole-dipole interacting cupric ions, in this case about 6 Å apart. The EPR absorption behaves according to the Curie (1/T) Law down to 14 K which indicates that exchange coupling is not significant. When it is necessary to work with small amounts of protein, and hence with small numbers of metal ions and poor signal-to-noise ratios, it may be difficult to know whether or not some cupric EPR absorption has been shifted to unexpected regions of field strength or so broadened that some of the integrated intensity is lost. There can, therefore, be doubt as to the number of truly EPR nondetectable copper ions.

The well-characterized enzymes fungal (*Polyporus versicolor*) laccase and tree (*Rhus vernicifera*) laccase have four copper atoms per molecule, one each of Types 1 and 2, and two that are EPR-nondetectable. Ferroxidase (ceruloplasmin) contains six to eight coppers of which approximately two are of Type 1 and two of Type 2, and two to four EPR-nondetectable. Ascorbate oxidase, with eight copper atoms, appears to have both Type 1 and Type 2 cupric sites among an estimated six which are EPR-detectable (Dawson, 1966). All of these blue oxidases have a strong absorption band at 30 kK as well as EPR-nondetectable copper. The presence of overlapping Type 1 and Type 2 absorption in the blue oxidases complicates the EPR spectra; nevertheless, a great deal of useful information about these enzymes has been obtained with this technique. First of all, the EPR spectra give no indication of interaction between cupric ions on the same molecule; the Type 1 and Type 2 sites must therefore be significantly separated (greater than 8 Å). The latter conclusion is supported by the result of a modification of fungal laccase; removal of the Type 2 cupric ion inactivates the enzyme but does not affect the spectroscopic properties of the remaining Type 1 ion (Malkin et al., 1969a). While none of the copper is very effective in relaxing water protons, an observation which suggests that all of the paramagnetic metal ions are "buried", the Type 2 cupric sites are accessible to small ligands as demonstrated in the investigations described below.

The enzyme activity of laccase is reversibly inhibited by fluoride and cyanide (Malkin et al., 1968; Malmström et al., 1968). Upon the addition of one equivalent of F^-, there is some change of g_z and A_z^{Cu} for the Type 2 site, but more striking is the ligand hyperfine splitting of the low field Type 2 copper hyperfine bands into doublets ($I^F = 1/2$). An additional equivalent of F^- will bind and the doublets become 1:2:1 triplets (equal coupling of the two fluoride nuclei to the cupric hole). These successive changes in the nonblue cupric site are not accompanied by alteration of the Type 1 absorption. There is unusual difficulty in removal of the second of the two coordinated fluoride ions, and also an experimentally useful change in reduction behavior, to be mentioned later, but otherwise the reaction with fluoride is simpler than the one with cyanide. Upon cyanide addition, the blue color bleaches and about 60% of the total EPR absorption is lost; the residual EPR spectrum is similar to spectra from cupric peptide complexes at alkaline pH which show resolved nitrogen hyperfine structure (earlier this chapter). The metal is linked directly to the carbon of CN^-, as can be seen by the appearance of carbon hyperfine coupling when ^{13}C-enriched cyanide is used. When the Type 2 copper is removed from laccase and the modified protein treated with cyanide, the resulting EPR absorption is less than 20% of, and does not resemble, the signal from the cyanide-treated native protein. Thus, it appears that the EPR spectrum of native laccase when cyanide is present arises from the nonblue (Type 2) cupric site, that cyanide binding occurs at this site, and that this binding profoundly affects both the properties of the Type 1 site and the enzymatic function of laccase. In a similar manner ferroxidase reacts with and is inhibited by azide, cyanide (concentration must be kept low to avoid removal of copper), and fluoride (Andréasson and Vänngård, 1970; Curzon, 1966; Kasper, 1968; Speyer and Curzon, 1968); binding to the Type 2 sites has been demonstrated or implicated in each case.

Magnetic susceptibility measurements which have been made on fungal laccase and ferroxidase indicate that the EPR-nondetectable coppers have little if any paramagnetism at room temperature (Ehrenberg et al., 1962). It was natural to assume these to be cuprous ions. However, a line of evidence has developed which might suggest that these EPR-nondetectable coppers are present as strongly coupled Cu(II) - Cu(II) pairs. First of all, in laccases and in ferroxidase, oxidation-reduction titrations show that as many electrons can be accepted as there are copper atoms (Carrico et al., 1971; Fee et al., 1969). In fungal laccase (four coppers per molecule) titrated anaerobically with octacyanotungstate, for example, bleaching of the band at 16 kK is proportional to reducing equivalents added and is complete when this number reaches 3.7, at which point no more oxidation of octacyanotungstate occurs. With other reducing agents, similar optical behavior is found, and the analogous magnetic measurements show the disappearance of all of the Type 1, but only part of the Type 2, EPR absorption when the number of reducing equivalents reaches four. Thus, there are approximately two electron-accepting sites more than the two present as single Type 1 and Type 2 sites. Kinetic studies both of anaerobic reduction and of reoxidation by molecular oxygen support this conclusion (Malmström et al., 1969). Secondly, a strong UV absorption band, at 30 kK, is associated with these additional reducible sites. In the anaerobic titration of fungal laccase, the decrease in absorption of the bands at 16 kK and 30 kK is the same linear function of added electron equivalents; in tree laccase, the decrease in absorption of the band at 30 kK proceeds somewhat in advance of the band at 16 kK. Clearly, the oxidation-reduction potential of the couple associated with the transition at 30 kK is comparably high with that of the couple involving the Type 1 site (about 0.77 volt for fungal laccase at pH 6.25; Fee and Malmström, 1968). Finally, no EPR absorption appears to be associated with these extra electron accepting sites in the oxidized, partially reduced, or fully reduced states (Fee et al., 1969; Malkin et al., 1969b); the electrons are received in pairs. (The latter experiments depend in part upon a curious property of fluoride-treated fungal laccase: reduction of the 30 kK band, which was in phase with that of the 16 kK band, now follows it; one electron bleaches the 16 kK band, two are then required to discharge the 30 kK absorption, and then finally the Type 2 EPR absorption declines.) These three features raise the interesting possibility that an electron-accepting role may belong to pairs of cupric ions interacting in some special manner, but there is at present no clear picture of what structure or mechanism may be involved.

Chapter 4

Heme Iron

1. Valence and Spin States of Iron

The common valence states and electron configurations of iron are ferrous, Fe(II)$3d^6$, and ferric, Fe(III)$3d^5$. Hemoglobin and myoglobin carry out their physiological roles with the metal remaining ferrous. Iron in the cytochromes cycles between ferrous and ferric. In the "resting" state, the iron in peroxidases and catalase is ferric. These enzymes gain one or two oxidizing equivalents after reacting with peroxides, and higher oxidation states for the iron in the catalytic intermediates have been discussed, Fe(IV)$3d^4$ in the form of the ferryl ion Fe(IV)O_2^{2+}, and Fe(V)$3d^3$. Compatible with the ferrous configuration are three spin states, S = 2,1,0 (diamagnetic) of which the first and last are found for deoxygenated and oxygenated hemoglobin (and myoglobin) respectively (Chap. 6-3). Similarly, three spin states are possible for the ferric configuration, S = 5/2 (high-spin), 3/2 (mid-spin), and 1/2 (low-spin). Loew (Harris, 1968a,b; Harris-Loew, 1970) has reviewed and extended the theory of spin state stability for the ferric ion in tetragonal complexes; the high- and low-spin states of ferric heme are greatly favored over the state of intermediate spin and over spin-mixed states. In ferrous hemoglobin and myoglobin, coordination numbers of five and six are associated with the high-spin (as in deoxy-hemoglobin) and low-spin (as in oxy-hemoglobin) states respectively. The ferric ion in hemeproteins is six-coordinated; changes in spin-state accompany changes in the sixth position, the ligand at that position producing a ligand field strength in accordance with its location in the spectrochemical series (Cotton and Wilkinson, 1972; Phillips and Williams, 1966). In small heme compounds (Hoard, 1968) and in hemeproteins (Chap. 2-2c), displacement of the iron from the heme plane is associated with the spin state changing from low to high.

2. Magnetic Susceptibility

Magnetic susceptibility, historically one of the first physical measurements to be made on respiratory pigments, occupies a central position in the characterization of transition metal ion systems, providing information about the average magnetic moment and hence about the spin state or states. It is the volume magnetic susceptibility κ of a medium (taken isotropic for the purposes of the immediate discussion) which is measured, κ being the constant of proportionality relating the magnetic polarization $\underline{M}$ within the medium to the applied field $\underline{H}$:

$$\underline{M} = \kappa \underline{H}.$$

Structural information is in the molar susceptibility χ which can be obtained from the measured volume susceptibility by

$$\chi = 10^3 \kappa / [M]$$

where [M] is the concentration in mol per liter and κ is in emu cm^{-3}.

To calculate the molar susceptibility from a quantum mechanical model, the energy levels (eigenvalues) E_j of the Hamiltonian are used to construct the partition function $Z(H,T)$:

$$Z(H,T) = \sum_j \exp[-E_j(H)/kT].$$

The Gibbs energy density of N paramagnetic ions per unit volume is then

$$E_V = -NkT \ln Z.$$

Noting that the magnetic energy density is $-\frac{1}{2}\kappa H^2$, one finds

$$\kappa = -\frac{1}{H}\frac{\partial E_V}{\partial H} = \frac{NkT}{H}\frac{\partial}{\partial H} \ln Z.$$

Also

$$\chi = \frac{N_{Av}kT}{H}\frac{\partial}{\partial H} \ln Z,$$

where N_{Av} is the Avogadro number. In general, for iron group transition metal ions at room temperature and below, to a good approximation one need only consider the ground electronic state. For states of spin $S = 1/2$, and for states in which $S > 1/2$ but with zero-field splittings small compared with kT, it can be shown that

$$\chi = N_{Av}\beta^2 g^2 S(S+1)/3kT$$

where β is the Bohr magneton and g is the g-value or spectroscopic splitting factor. The molecular magnetic dipole moment μ is then $g\beta\sqrt{S(S+1)}$, (e.g. in high-spin Fe(III), $g = 2.0$ for $S = 5/2$ manifold). This situation is schematized in Figure 4-1. When the zero-field splitting is appreciable, one most employ the partition function of the system. For example, in high-spin ferric heme compounds, where the zero-field splitting parameter D is in the range 5-15 cm^{-1}, $kT < D$ at liquid helium temperature. (The thermal energy corresponding to 1°K is 0.7 cm^{-1}; liquid helium boils at 4.2 K.) The relative energies in this case are shown in Figure 4-2, and the corresponding calculation will now be outlined.

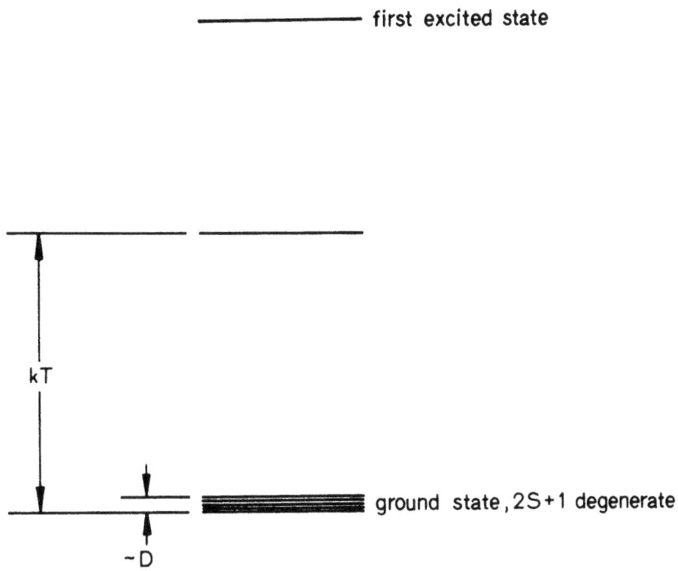

Fig. 4-1. Energy level diagram for the case where thermal energy is large compared with the ground state zero-field splittings and small compared with excited state energies

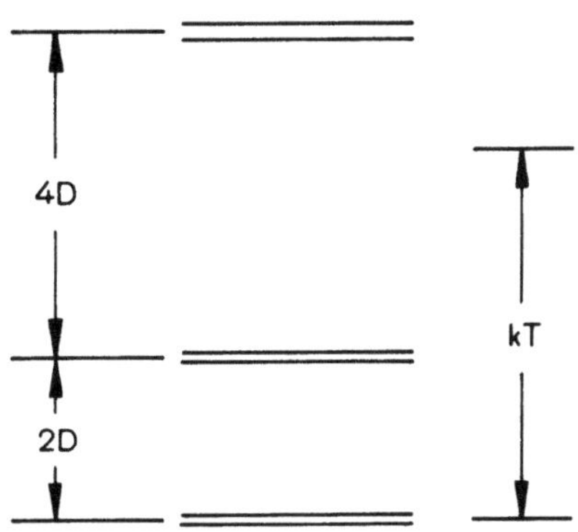

Fig. 4-2. Energy level diagram for the case where thermal energy is comparable with ground state zero-field splittings. The $^6A_1$ ground state for the high-spin ferric ion is shown

Consider a collection of heme sites for which the applied magnetic field is perpendicular to the heme plane.

$$Z(H,T) = e^{x} + e^{-x} + e^{-y+3x} + e^{-y-3x} + e^{-3y+5x} + e^{-3y-5x} + e^{-z+x} + \ldots$$

where

$$x \equiv \beta H/kT,\ y \equiv 2D/kT, \text{ and } z \equiv \left\{E\left(^4A_2\right) - E\left(^6A_1\right)\right\}/kT,$$

$^4A_2$ being the first excited electronic state.

Since

$$E\left({}^4A_2\right) - E\left({}^6A_1\right) \approx 4000\ \text{cm}^{-1},\ z >> x,y,$$

it follows that

$$Z(H,T) = \sum_{n=1}^{3} e^{n(1-n)y}\left\{e^{(2n-1)x} + e^{(1-2n)x}\right\}.$$

For $\beta H < kT < 2D$, one finds then that the molecular susceptibility normal to the heme plane is approximately β^2/kT. In the same way, the molecular susceptibility in the heme plane can be shown to be about $9\beta^2/kT$. These are the values which would be found from single crystal measurements at very low temperatures; frozen solutions and polycrystalline samples would have an isotropic molecular susceptibility of $(1+2\times 9)\beta^2/3\,kT$ at these temperatures, corresponding to a dipole moment of $\sqrt{19}\ \beta$, as compared with a moment of $\sqrt{35}\ \beta$ for room temperature solutions. The temperature dependence of the susceptibility can be calculated by the method just given and D adjusted to fit the experimental data (Uenoyama et al., 1968). Other reliable methods of determining the zero-field splitting include the direct measurement by far-infrared spectroscopy (Brackett et al., 1971; Richards et al., 1967) and an indirect technique based upon the temperature dependence of the spin-lattice relaxation time (Scholes et al., 1971). When a wide range of wavelengths, down to submillimeter, is available for EPR, the frequency dependence of $g_\perp$ can be used to determine D (Alpert et al., 1973).

The discussion above dealt with the paramagnetism of a spin-system structurally independent of temperature changes when subject to moderate magnetic fields. Spin state equilibria will be considered later (Chap. 4-5) and the reader is referred elsewhere for experimental information (Brill, 1969) and treatment of strong-field effects (Loew, 1972).

3. Valence State Determination

In general, freezing will not alter valence, and when the EPR measurement at cryogenic temperature is available, the presence of the ferric state is readily ascertained if the high-spin spectrum, with the characteristic $g = 6$ resonance, appears. EPR spectra of the kind exhibited by low-spin ferric heme compounds could conceivably arise from other configurations but are, nevertheless, very good evidence for the +3 state. In the absence of a ferric EPR spectrum the most likely valence for the iron is ferrous. (We assume here that the metal ions are not strongly spin-coupled.)

A diamagnetic susceptibility for the metal ions would in itself indicate the (low-spin) ferrous state since d^5 (and d^3) configurations must have nonzero paramagnetic moments and d^4 electrons would not all be paired (except under most unusual conditions). While susceptibility measurements yielding unusual paramagnetic moments have been used in speculations about special valence in the compounds studied [e.g. ~ 3 β in support of the low-spin ferryl ion; ~ 4 β in support of Fe(V)], such data are equivocal, and EPR spectroscopy of these systems has not been productive of information about the iron. The measurement of magnetic susceptibility offers a reliable diagnostic of valence and spin state only for high-spin ferric and low-spin ferrous.

In principle, Mössbauer spectroscopy offers the possibility of distinguishing the valence states of iron. The energy of the transition between the ground state ($I = 1/2$) and 14.4 KeV excited state ($I = 3/2$) of the ^{57}Fe nucleus used as a Mössbauer probe depends in part upon the total charge density of s-electrons at the nucleus (acting through a small difference in nuclear radius between the ground and excited states). The s-electron density depends upon the electron configuration and, for essentially ionic compounds Mössbauer spectra exhibit "chemical" or "isomer" shifts which are characteristic of the valence state (ferrous and ferric differing by about 0.9 mm/sec). However, the s-electron density also depends upon electron distribution and is strongly influenced by bond formation, both through the admixture of 4s orbitals into metal levels of lower energy and through delocalization (metal electrons onto the ligands and conversely). For example, in the case of the low-spin complexes $K_4Fe(II)(CN)_6 \cdot 3H_2O$ and $K_3Fe(III)(CN)_6$, there is essentially no difference in isomer shift between the ferrous and ferric compounds; Shulman and Sugano (1965) have shown that the delocalizations required to produce the observed g-values (it is necessary to use EPR data from the Mn(II) complex for the Fe(II) case) are such that the effective charges on the metal ions in the two valence states are approximately the same. The converse situation also occurs, where appreciably different isomer shifts are found for two complexes of iron in one particular valence state; of special interest in this connection is ferrous hemoglobin which exhibits isomer shifts differing by 0.65 mm/sec between the deoxy- and oxy-forms (Lang and Marshall, 1966). Clearly, the isomer shift does not in itself provide an unambiguous answer to the question of metal valence. Additional information related to electron configuration is available in the comparison of Mössbauer spectra in the presence and absence of an applied magnetic field, but this data does not uniquely determine the iron valence state.

Thus, the electron configuration of the iron in a heme compound in some cases cannot be recognized with certainty on the basis of the body of knowledge available at the present time. Usually the subject compound can be oxidized or reduced to a state for which the configuration is determinable (e.g. one which gives the high-spin ferric EPR spectrum). The question then arises as to whether the electron transfer has involved the iron or an organic group, or both if more than one equivalent has been passed. As discussed in the following section, optical spectroscopy can be helpful in answering this question, but obscurity remains in some cases.

4. Optical Properties

In the wavelength region between 300 and 700 nm, the absorption of light by hemeproteins is associated with excitation of the porphyrin structure, the states of which are influenced by the iron (and hence by the ligands in positions 5 and 6 of the metal ion) and by the protein. There is often significant absorption between 700 and 1700 nm which is not present in the metal-free porphyrin. From 200 to 300 nm, almost all of the absorption is due to chromophores of the protein moiety, but heme does have bands in this region; changes in the state of the iron perturb the latter bands and some of the bands associated with protein groups as well.

Roughly speaking, the band at about 400 nm (Soret band) has a maximum molar absorptivity of the order of 10^5 M^{-1} cm^{-1}, the bands between 450 and 650 nm, 10^4 M^{-1} cm^{-1}, and the bands between 700 and 1000 nm, 10^3 $M^{-1}cm^{-1}$. Thus, there is a ten-fold drop in intensity as the wavelength increases from the near-ultraviolet to the visible, and another ten-fold drop in going from the visible to the near-infrared. It follows that an extremely large range of concentrations can be accomodated in optical absorption spectrophotometric determinations of hemeprotein concentration and in detection of changes in heme electronic structure. There is a rich literature (the references given below are only representative) of experimental data, empirical correlations, and quantum mechanical explanations of the energy levels of porphyrin molecules (Falk, 1964; Gouterman, 1961; Kobayashi, 1959; Platt, 1956; Simpson, 1948; Weiss et al., 1965) and of the influences of metal valence, ligands and spin-state, porphyrin and protein structure, and solvent perturbations on the optical absorption spectra of hemes and hemeproteins (Antonini and Brunori, 1971; Braterman et al., 1964; Brill, 1966; Brill and Sandberg, 1968; Brill and Williams, 1961; Chance, 1947; Chance and Schonbaum, 1963; Eaton and Charney, 1969; Eaton and Hochstrasser, 1967; George et al., 1961; Hartree, 1946; Haurowitz et al., 1935; Keilin and Hartree, 1935; Lemberg and Legge, 1949; Margoliash and Schejter, 1966; Offenhartz, 1965; Scheler et al., 1957; Smith and Williams, 1970; Stellwagen et al., 1967; Theorell, 1942; Theorell and Ehrenberg, 1951; Williams, 1956; Zerner et al., 1966).

One seeks first a structural feature common to all porphyrins as responsible for the strong near UV absorption (Soret or "B" transitions) and the weaker absorption in the visible ("Q" transitions). In the macrocyclic tetrapyrrole arrangement of porphyrins there are a 20 carbon atom outer ring and a 16 atom (four are nitrogens) inner ring. These two rings share four methene bridges. The aromatic character of porphyrins is shown by their heats of combustion and by the bond lengths determined with high resolution X-ray diffraction. Furthermore, porphyrin structures arrived at by the latter method reveal that the greater part of the *delocalized* π bonding is confined to the inner 16-membered ring (Hoard, 1968). Two of the nitrogens in this ring are of the pyridine type (Chap. 6-4), each contributing one electron to the delocalized system (as do each of the 12 carbons), and two are pyrrole nitrogens, each contributing two electrons to the π system. Hence there are 18 electrons in a delocalized system extending over the inner porphyrin ring. In the lowest order approximation to this system, which restricts noninteracting electrons to a circular ring on which the potential energy is constant, the single electron wave-functions are of the form $\psi_m = \frac{1}{\sqrt{2\pi}} e^{\pm im\phi}$ with energies $E_m = \frac{1}{2}\frac{m^2\hbar^2}{m_e r^2}$, where ϕ is the angle measured around a circle of radius r and m is zero or any positive or negative integer. The filled orbitals of highest energy are those with $m = \pm 4$, and the lowest lying empty orbitals are those with $m = \pm 5$. There are then four molecular terms, each orbitally two-fold degenerate, which arise out of the lowest energy one-electron transitions and involve the four orbitals characterized by $m = \pm 4$ and $m = \pm 5$: (1) a spin singlet with total orbital angular momentum $M = \pm 1$, called ^1B after Platt (1956); (2) a spin singlet with $M = \pm 9$, called ^1Q; (3) and (4) the corresponding spin triplets ^3B and ^3Q. The transitions to the triplet states are spin-forbidden, the ground state being a singlet. Furthermore, the ground state has $M = 0$, and so the transitions to the B states are both x ($r\cos\phi = \frac{r}{2}(e^{i\phi} + e^{-i\phi})$) and y ($r\sin\phi = \frac{r}{2i}(e^{i\phi} - e^{-i\phi})$) allowed, while the transitions to the Q states are nominally forbidden but do occur with less strength. Upon inclusion of electron-electron interactions, the generalization of Hund's second rule (Chap. 6-3) to the orbital angular momentum states of molecules places Q states lower in energy than the B states. As mentioned in Chapter 6-5, different vibrational levels of the Q states give rise to two bands; in metalloporphyrins one of these is usually in the region 600-560 nm (the α band) and the other, 560-520 nm (the β band). In this manner the 18 electron inner ring model exhibits the basic spectral properties of porphyrins in the visible and near UV wavelength regions. Support for this model comes from measurements of the Zeeman effect and magnetic circular dichroism, to be discussed immediately below. Furthermore the results of self-consistent molecular orbital calculations (see, e.g., Weiss et al., 1965) are in agreement with the inner ring being the main path of delocalization and provide an explanation of spectral bands based upon a four-orbital model like the simple one just given.

We take up now magneto-optical properties of porphyrin. In the 18 electron ring model, the ground state is diamagnetic ($M, S = 0$) and the B and Q electronic excited states have $M = \pm 1$ and ± 9 respectively. Focusing upon the singlet Q states, for example, and subjecting the molecule to a magnetic field perpendicular to the ring, we expect the energies to be shifted by $M\beta H$ and thus there to be a Zeeman splitting of $\Delta E = 2|M|\beta H$ or about 82 cm^{-1} for a 100 KG field. In solution, the molecules are randomly oriented with respect to the applied field, halving the Zeeman effect so that 100 KG would produce a splitting of only about 41 cm^{-1} or 13 Å for a band at 5700 Å. The Q bands are many times broader than this splitting, and it is necessary to use left and right circularly polarized light in order to observe separately the peak wavelengths of the transitions to the +M and -M states (Feher et al., 1967). The difference in absorptivity between left (LCPL) and right (RCPL) circularly polarized light as a function of the strength of the applied magnetic field is more often measured directly and the term magnetic circular dichroism (MCD) is usually reserved for this difference measurement. In principle, both the difference and absolute spectra can be employed to determine the angular momenta and splitting of nearly degenerate excited states; however, asymmetries between the LCPL and RCPL absorption curves provide more sensitive detection of nondegeneracy while the MCD difference spectrum gives a better estimate of the angular momentum (Sutherland et al., 1971). Closely related to this Zeeman phenomenon is the Faraday effect or magneto-optical rotatory dispersion (MORD) wherein the plane of polarization of linearly polarized light is rotated on passing through material subjected to a magnetic field along the direction of the light beam. The use of MORD to study porphyrin derivatives (Shashoua, 1964, 1965) preceded that of MCD, but the latter has been more widely used.

Realistic interpretation of MCD spectra requires a more detailed description of porphyrin excited states than that provided by electrons moving freely around a circle. Focusing once again upon the Q transitions, we find that in general there are not two degenerate members of a Q state (which would have $L_Z = 9\,\hbar$ and $-9\,\hbar$ in a magnetic field) but rather there are non-degenerate states Q_x and Q_y to which distinct but (usually) unresolved transitions occur, polarized at right angles in the molecular (xy) plane. The separation in energy of Q_x and Q_y can be viewed as arising from a spatial perturbation, such as $V = \Sigma v_n \cos n\phi$, which in the absence of a magnetic field leads to orbital singlets

$$|Q_x\rangle = N\left\{\frac{1}{\sqrt{2}}(|M\rangle + |-M\rangle) + a(|M-1\rangle + |-M+1\rangle) + b(|M+1\rangle + |-M-1\rangle) + \ldots\right\}$$

and

$$|Q_y\rangle = iN\left\{\frac{1}{\sqrt{2}}(|M\rangle - |-M\rangle) + a(|M-1\rangle - |-M+1\rangle) + b(|M+1\rangle - |-M-1\rangle) + \ldots\right\}$$

where $N < 1$ is a normalization coefficient, $\langle Q_y|L_Z|Q_x\rangle = -iM'\hbar$, M' is usually less than $M = 9$, and the levels are separated by a zero-field splitting $2\,\Delta E_{ZF}$. A z-directed magnetic field will now mix $|Q_x\rangle$ and $|Q_y\rangle$ to form states with energies diverging as $2\sqrt{(\Delta E_{ZF})^2 + (\beta M' H)^2}$. If ΔE_{ZF} is small compared with the width of the absorption bands, the assumption of degeneracy (neglect of ΔE_{ZF}) leads to little error in the calculation of M' from MCD and zero-field absorption spectra (e.g. by the method of Stephens et al., 1966). It is necessary to know ΔE_{ZF} when it is appreciable, and also the zero-field absorptivities for the x and y transitions separately and the shape and width of these bands, in order to be able to determine M' from the MCD. Sutherland and Klein (1972) have developed this analysis and applied it to the 550 nm band of ferrocytochrome c for which they find $\Delta E_{ZF} = 50 \pm 5\ cm^{-1}$ and $M' = 7.9$ (at room temperature). These parameters are consistent with near degeneracy (the free electron ring system appears to be a good model for porphyrin transitions in ferrocytochrome c) and essentially the same value of M' is obtained when ΔE_{ZF} is taken to be zero. For a few metalloporphyrins M' is as low as four; and where LCPL and/or RCPL absorption curves are available, the asymmetries in these suggest that ΔE_{ZF} often exceeds 100 cm^{-1}. The value of M^{l} for certain metalloporphyrins has been shown to change in going from room to cryogenic temperatures, but in no case has the structural basis for the removal of the degeneracy of the excited states been established. It should be emphasized here that the generally large values (six to eight) of M^{l} which are found experimentally support the molecular orbital picture derived on the basis of axial symmetry, and that the departure from degeneracy of the members of successive pairs of states is small compared with the energy differences between pairs.

The MCD bands just discussed, which arise from the Zeeman splitting of degenerate excited states, are "A" terms. More generally an MCD "A" term is one which has its origin in the removal of the degeneracy of a ground or excited state by a magnetic field. There are also "B" terms, which reflect the mixing of the ground or excited state with other excited states, and "C" terms, which arise from the unequal (Boltzmann) distribution of population among the members of a degenerate ground state split by a magnetic field. The latter bands increase in intensity approximately as the inverse temperature. For $\pi-\pi^*$ (B and Q) transitions of porphyrin the ground state is non-degenerate, and C terms are not associated with them. In paramagnetic metalloporphyrins these transitions could have C character to the extent that the metal ion is involved. While this is not the case for most non-iron metalloporphyrins, the paramagnetic states of myoglobin all have C terms in the $\pi-\pi^*$ regions of the MCD spectra, and several features of the spectra can be correlated with heme spin state (Vickery et al., 1976a; Springall et al., 1976). In Figure 4-3 are shown myoglobin MCD and absorptivity spectra, in the Soret and visible regions, for complexes representative of the four common valence/spin states of heme iron. Where data is available from other hemeproteins, the MCD behavior is similar to that of myoglobin.

Absorption spectra from microscopically thin hemeprotein single crystals taken with plane polarized light (Kendrew and Parrish, 1956; Perutz, 1953) also exhibit features associated with the ring model for porphyrin. Eaton and Hochstrasser (1967, 1968) introduced the concept of polarization ratio (PR) spectrum in connection with studies of this kind. Consider, for example, horse heart ferricytochrome-c crystals. In these tetragonal crystals the c-axis is a four-fold rotation axis relating the four protein molecules per unit cell, and the a- and b-axes (perpendicular to c) are equivalent. The polarization ratio is then

$$PR \equiv \frac{\text{absorption for light polarized along the crystal c-axis}}{\text{absorption for light polarized along the crystal a-axis}}$$

We will define the z-axis to be the normal to the heme plane, and x and y to be in the heme plane, as before, but whatever one takes for the molecular x, y, z axes, each of these will then make the same angle with the c-axis for every molecule. The absorption is proportional to the square of the projection of the transition dipole moment upon the electric vector of the plane polarized light, and so, for a transition polarized normal to the heme plane,

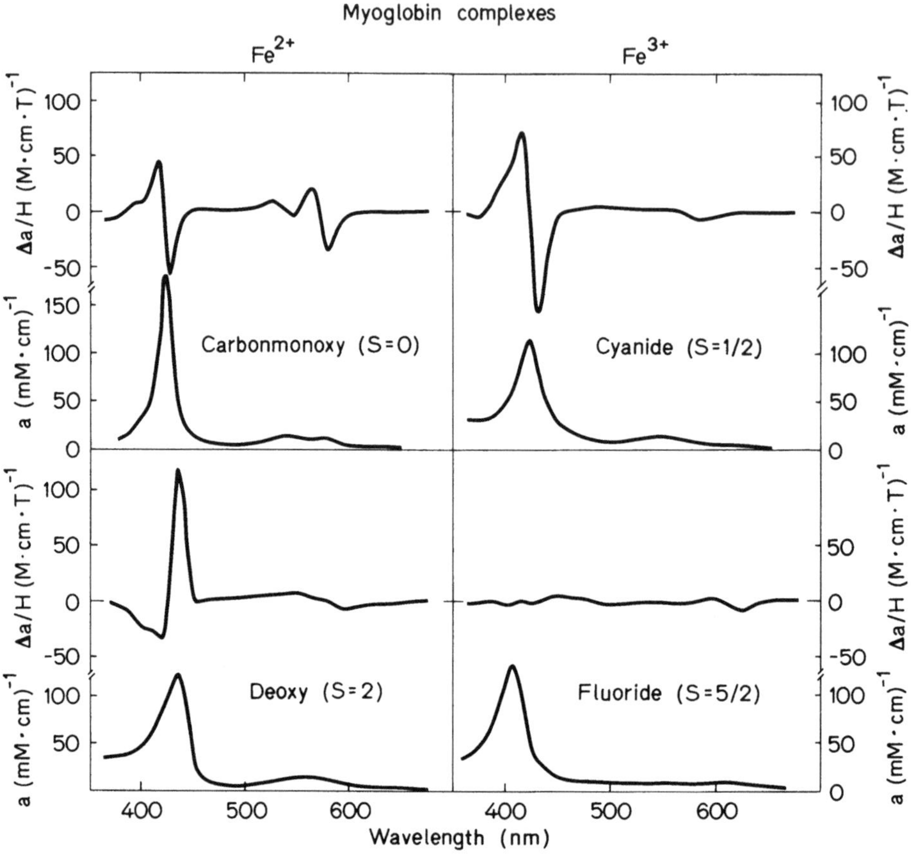

Fig. 4-3. Room temperature MCD and absorptivity spectra of four myoglobin complexes. T = 10,000 G (Vickery, 1978)

$$PR(z) = \frac{4\cos^2\alpha}{\sin^2\alpha[\cos^2\beta + \cos^2(\beta+\frac{\pi}{2}) + \cos^2(\beta+\pi) + \cos^2(\beta-\frac{\pi}{2})]}$$

$$= 2\cos^2\alpha/\sin^2\alpha\,,$$

where α is the angle the heme normals (z_1 through z_4) make with the c-axis and β is the angle which the projection of heme normal z_1 onto the ab plane makes with the a-axis. Similarly, for transitions polarized isotropically in the heme (xy) plane,

$$PR(x,y) = \frac{4\int_0^{\pi}\sin^2\alpha\cos^2\phi\,d\phi}{2\int_0^{\pi}\cos^2\alpha\cos^2\phi\,d\phi + 2\int_0^{\pi}(1)^2\cos^2\phi\,d\phi}$$

$$= 2\sin^2\alpha/(\cos^2\alpha + 1)\,,$$

where the integral is in the xy plane. In these crystals of cytochrome c, the angle α is 70° (Dickerson et al., 1967), and one finds PR(z) = 0.26, PR(x,y) = 1.58. In this case, the PR provides a critical measure of the direction of the transition moment, which for B and Q transitions would be in the heme plane. Eaton and Hochstrasser (1967) find that the PR spectrum is relatively constant from 380 to 540 nm with a value of 1.7, confirming that the Soret and main visible band (530 nm) transitions are polarized in-plane. These data also require the absorption in the porphyrin plane to be essentially isotropic; that is, as predicted, the Q_x and Q_y states are almost equivalent (degenerate). As discussed in the preceding section, some splitting of Q_x and Q_y often occurs, and the effect of this is observed in the asymmetry of the absorption spectra taken with LCPL and RCPL. Departure, near the band center, of the PR spectrum from the perfectly isotropic value would be related to the latter. Such an effect is seen at the Soret band, where it is very small but reproducible, and at the band at 565 nm where it is substantial, reaching a positive value of 2.4 at about 580 nm. The band at 565 nm is polarized in-plane, but is not isotropic in the plane - Q_x and Q_y differ significantly. The band at 695 nm, which has a PR of less than 0.5, is clearly recognized as being different in kind; when the residual intensity from x,y absorption is subtracted out, the PR approaches that for a z-polarized transition.

Insofar as they relate to the basic electronic structure of the porphyrin system, the same features of linearly polarized light single crystal absorption spectra have been found for ferrimyoglobin complexes as for ferricytochrome c (Eaton and Hochstrasser, 1968). Additional information is obtained for the high-spin complexes which, in the region from 450 to 650 nm, have four bands; of these bands two (pairs) are different vibronic components of the porphyrin Q transitions and the other two are generally viewed as (porphyrin to iron) charge-transfer transitions. The PR spectrum for the ferrimyoglobin high-spin complexes is nearly constant throughout the region of these four bands and has a value appropriate to x,y polarization. Thus, like the Q states, the charge-transfer states are spatially degenerate, which implies that the iron d_{xz}, d_{yz} orbitals are the ones involved in the charge-transfer transitions.

While the PR reflects only the directional properties of transitions, the measured absorbances for light polarized parallel to the crystal axes can be used (in conjunction with values of the crystal thickness and heme concentration) to calculate the absorptivity which a randomly oriented system of molecules would have. For horse-heart ferricytochrome c and for ferrimyoglobin complexes, there is good agreement between the isotropic absorptivity calculated in this way and the value obtained from solution measurements in the wavelength region from 400 to 700 nm (Eaton and Hochstrasser, 1967, 1968). At longer wavelengths (700-1600 nm), Day et al. (1967a) find that ferricytochrome c, ferrimyoglobin complexes, and cytochrome c peroxidase all show a reduction in the intensity of absorption bands in the crystalline state as compared with solution.

The natural optical activity which hemeproteins exhibit in the near UV and visible is not an intrinsic property of porphyrin structure. As discussed above for porphyrin, the $\pi-\pi^*$ electric dipole transitions of planar aromatic molecules are polarized in-plane. For reasons of symmetry, there cannot be the corresponding in-plane magnetic dipole moment necessary for optical activity. Coupling to d-d transitions of the iron does not provide a major part of the magnetic dipole character; this follows from the fact that complexes of porphyrins and apohemoglobin have CD spectra with essential features similar to those of native hemoglobin (Ruckpaul et al., 1970; Sugita et al., 1971). One then looks for rotational strength arising from coupling of the porphyrin (heme) transitions with excited states of the protein side chains and backbone, and, possibly, from distortion of the porphyrin ring lowering the symmetry sufficiently to make it optically active. Hsu and Woody (1971) investigated theoretically the sources of the natural optical activity of the near UV and visible bands in myoglobin and hemoglobin and found that the major contribution comes from interactions between heme transitions and those of nearby aromatic side chains. This implies that the local arrangement of aromatic residues, and hence the polypeptide conformation, will affect the CD spectrum, a view supported, for example, by observed differences in CD spectra between the α and β chains of human hemoglobin, between hemoglobin chains and myoglobin, and among hemoglobins from divergent species (Sugita et al., 1971).

The differences in the origin of MCD and natural CD in the near UV and visible bands of hemeproteins should be noted, MCD being an intrinsic property of porphyrin and CD requiring the participation of protein groups. The MCD spectrum is most influenced by structural changes within the coordination sphere of the iron which affect the ground and lowest-lying excited states of the heme group. The CD spectrum is most influenced by changes in the protein surrounding the heme. A comparison, for example, of the absorption, MCD, and CD spectra in the Soret region of ferricytochrome c and its heme undecapeptide (in the presence of 3.3 M N-acetyl-DL-methionine to provide a methionine sulfur for the sixth iron ligand) shows that the absorption and MCD spectra do not differ between the protein and its fragment whereas the CD spectra are different (Risler and Groudinsky, 1973). Of course, change in metal valence is accompanied by changes in the separation and width of the porphyrin energy levels, and usually by some change in protein conformation as well; so it is not surprising to find significant valence-dependence of hemeprotein optical activity (Drucker et al., 1970; Ulmer, 1965; Urry and Doty, 1965). One also expects that spin-state conversion, which is accompanied by changes in the absorption spectrum and by a displacement of the iron atom which causes local structural movement, would be attended by changes in natural optical activity; this is the case, for example, when carbon monoxide and molecular oxygen complexes are formed from the free ferrous hemeproteins (Sugita et al., 1971; Willick et al., 1969).

As the radiation goes from the Soret plus visible region to the near infrared, the dominant roles in the absorption of light by heme systems pass from the $\pi-\pi^*$ B and Q transitions to charge transfer and d-d transitions. Since the metal ion is more intimately involved in transitions of the latter two types than in

π-π* absorption, one expects the near IR to be more productive of information about the iron; but exploration of this region has proceeded relatively slowly, partly because of the feebleness of the bands and hence the high concentration of protein required. In general, low-spin ferric complexes of hemeproteins in solution have absorptivities which are under 200 M^{-1} cm^{-1} from 750 to 950 nm and peak at 1000 to 1100 nm with values of about 300 M^{-1} cm^{-1}. The high-spin ferric complexes have two bands in this region, a weaker one at, roughly, 750 nm, and one with an absorptivity maximum of approximately 1000 M^{-1} cm^{-1} at, roughly, 850 nm (Brill and Sandberg, 1968; George et al., 1961). The location of the latter pair of bands varies considerably with the nature of the axial ligand; these bands are expected to involve charge transfer states since d-d transitions are spin-forbidden (as well as space-forbidden) in the high-spin ferric case. (No exceptions to the pattern of near IR bands described in this paragraph have been found for ferric hemeproteins; there are, however, unpublished reports of near IR data from simpler ferric heme systems which do not follow this pattern.) The absorption of light by ferrous hemeproteins is similarly weak (less than 300 M^{-1} cm^{-1}) and diffuse between 700 and 1000 nm. Based upon a combination of absorption and CD spectroscopy, Eaton and Charney (1969) place three greatly overlapping bands of ferrocytochrome c at 630 nm (250 M^{-1} cm^{-1}), 730 nm (180 M^{-1} cm^{-1}), and 820 nm (200 M^{-1} cm^{-1}, degenerate), and are thereby able to estimate the octahedral crystal field splitting (10 Dq) as 17,000 to 19,000 cm^{-1}. At wavelengths between 1000 and 1500 nm, the absorption by H_2O can be troublesome. While the use of buffers made up with D_2O for hemeprotein optical studies has thus far been limited, the results have been helpful both in assigning certain features of spectra taken in H_2O to the solvent and in reducing solvent absorption. Absorption spectra from single crystals of ferricytochrome c, ferrimyoglobin and ferrimyoglobin-cyanide have been taken with plane polarized light from 600 to 1600 nm (Day et al., 1967a,b); the low- and high-spin complexes exhibit a similar pattern of bands and all the bands are polarized in the heme plane. Measurements of MCD in the near IR (instrumentation for this has only recently become available) upon ferric myoglobin and hemoglobin compounds reflect differences between the low- and high-spin complexes (Cheng et al., 1973). Without magnetic fields, no CD is observed from the cyanide complex of ferrihemoglobin (in D_2O buffer) in the region 800 to 2000 nm wherein there are absorption maxima at 1080 nm (1000 M^{-1} cm^{-1}), 1280 nm (2200 $M^{-1}cm^{-1}$), and 1560 nm (4200 M^{-1} cm^{-1}). When a magnetic field is applied in the direction of the light beam, an MCD spectrum is observed which parallels the absorption spectrum with $(A_{LCPL} - A_{RCPL})/HA \approx 2 \times 10^{-4}(KG)^{-1}$. This data has been discussed in terms of paramagnetic C terms although no temperature dependence of the MCD was reported. The dependence of the MCD amplitude upon magnetic field strength was estimated on the basis of (charge transfer) excitation from porphyrin orbitals to the d-shell t_{2g} hole of the low-spin ferric ion and found to be in approximate agreement with the observed value. The CD of high-spin ferric myoglobin and hemoglobin was not detectable, and the MCD was large but different in shape from that of the low-spin cyanide complexes. Considerably more information from MCD studies in the near IR is likely to become available during the coming years. The discussion turns now to more detailed spectral effects accompanying structural changes.

Oxidation-reduction reactions of heme can involve the metal ion alone, the porphyrin ring, or both. Several spectral changes consistently accompany particular structural changes induced by electron transfer:

1. Reduction from ferric to ferrous increases and sharpens (especially when the ferrous conformation is low-spin) the peak absorptivities of the bands (α and β) between 500 and 600 nm, decreasing the absorption for some 50 nm above and below this region. Since the MCD maximum is inversely proportional to the square (approximately) of the band width, it magnifies these absorptivity changes, as in the case of ferricytochrome c versus ferrocytochrome c (Sutherland and Klein, 1972). (It has been suggested that the bandwidths - apart from unresolved splittings as in the Soret - of π transitions in metalloporphyrins may be related to the extent of mixing of porphyrin π and metal d_{xz} and d_{yz} orbitals, a mixing which would be small in the low-spin ferrous case where the t_{2g} orbitals are of low energy and filled; Hochstrasser, 1971).

2. Interruption of the porphyrin conjugated ring system (for example, by oxidation at a methene bridge carbon) greatly decreases the Soret absorptivity and introduces a new band at about 650 nm. The resulting solutions are green as in Compounds I of peroxidase and catalase.

3. Oxidations of ferric hemeproteins (brown in the high-spin state) that leave the porphyrin ring intact result in red Compounds "II" and "III". Compounds II have α and β bands at relatively short wavelengths (about 560 and 530 nm respectively). The optical spectra of Compounds III are somewhat similar to those of oxy-, carbonmonoxy-, and NO (ferrous) hemoglobin and myoglobin.

There are many reactions of hemeproteins which do not involve electron-transfer, such as the formation of complexes of the iron with small molecules at the sixth coordination position. The oxygenation of ferrous hemoglobin (and myoglobin) is in this category, a process accompanied by well-known changes in coordination number, spin-state, and color (absorption spectrum). In ferric hemeproteins, if the metal ion can bind a small ligand in the sixth position, this position is always found to be occupied. Reactions at this position are therefore substitutions (e.g. cyanide for water) or ligand ionization (e.g. deprotonation of water). The effects of substitution upon the absorption spectrum and magnetic moment parallel one another, being monotonically related to the strength of the complexing group in producing ligand-field splitting. Typically fluoride complexes of ferric hemeproteins have the high-spin magnetic moment of 5.9 Bohr magnetons, a Soret absorptivity maximum at 403-407 nm, and α and β bands of low intensity discernible between two other well-defined bands peaking at about 490 nm and 610 nm; the cyanide complexes have magnetic moments of 2.0-2.7 Bohr magnetons (the spin-only low-spin value is 1.73, but orbital angular momentum and, at room temperature, the population of excited states make significant contributions), Soret peaks at 419-425 nm, and just two visible bands, a relatively intense β band centered at about 540 nm and the weaker α band at about 575 nm. Thus, for example, as substitution through the series fluoride, water (pH 6.4), hydroxide, azide, cyanide proceeds, the magnetic moment of horse ferrimyoglobin takes on the values 5.9, 5.7, 5.1, 3.3, 2.0 and the Soret absorptivity maximizes at 406, 408, 411, 420, 422 (Scheler et al., 1957). Differences in heme binding site among proteins result in differences in the effectiveness of a particular molecule in producing ligand-field splitting; the azide complex of bacterial catalase has a magnetic moment of 5.4, while that of horse ferrihemoglobin has 2.4, the Soret bands peaking at 413 and 417 nm respectively.

The variation of pH alone can produce changes of spin states. For example, ferric hemoglobin and myoglobin exhibit a transition from high to intermediate spin as the pH is raised from 7 to 10 (Coryell et al., 1937; Theorell and Ehrenberg, 1951), and in cytochrome c the spin state of the ferric ion goes from high to low upon titration from pH 2 to 7 (Boeri et al., 1953). These magnetic susceptibility changes are paralleled by the characteristic changes in light absorption described earlier in this section, and magnetically-induced optical activity also shows the same titration behavior (Rein et al., 1973). With regard to the intermediate spin of alkaline ferric hemoglobin and myoglobin, evidence provided by the temperature dependence of the magnetic moments and absorption spectra, by a quantitative correlation between the moments and spectra, and by the effect of D_2O upon the visible absorption spectrum, strongly favors the existence of a thermal mixture of high- and low-spin complexes (George et al., 1961). Pressures of several thousand kg/cm^2 induce spectral shifts in ferric hemoglobin and myoglobin complexes consistent with transition from high- to low-spin (Fabry and Hunt, 1968; Ogunmola et al., 1976; Zipp et al., 1972). Spin state equilibria are discussed further in the next section (Chap. 4-5).

The spectral properties discussed above are associated with transitions which involve mainly porphyrin and metal orbitals. The absorption bands attributable to amino acid residues are large in number and overlap heavily so that it is difficult to resolve and assign spectral changes. In the region from 300 to 250 nm, where the transitions belong to the aromatic residues phenylalanine, tyrosine, and tryptophan, the absorption spectra are (other than the tryptophan notch at about 290 nm) rather insensitive to changes in valence and spin-state of the iron. In the region below 250 nm, where many residues absorb, difference spectroscopy reveals an increase in absorptivity (of the order of 10^4 M^{-1} cm^{-1}) when the spin state changes from high to low in ferric hemeproteins, and upon formation of oxidized compounds from ferric hemeproteins (Brill and Sandberg, 1967). Apart from porphyrin (and cyanide, if present) to metal charge transfer transitions in the region 230 to 210 nm, there is a difference band at about 240 nm which is believed to be due to perturbation of the imidazole ring of the proximal (coordinated) histidine residue. The latter assignment is supported by the results of absorption spectrophotometric studies of model systems such as heme-polyamino acid complexes (Sandberg and Balegh, 1973).

5. Spin State Equilibria

The possibility that high- and low-spin structures can be in thermal equilibrium was mentioned in the preceding section. One considers the measurements to be restricted to a range of protein concentrations in which the equilibrium is independent of the concentration and hence can be expressed

$$\frac{\alpha}{1-\alpha} = K,$$

where α is the fraction of the magnetic centers in the low-spin form and the equilibrium constant K is a function of temperature. The average magnetic moment $\bar{\mu}$ is then given by

$$\bar{\mu}^2 = \alpha \mu_L^2 + (1-\alpha)\mu_H^2$$

and the average absorptivity $\bar{a}$ by

$$\bar{a} = \alpha a_L + (1-\alpha) a_H$$

where the subscripts H and L refer to the high- and low-spin parameters respectively. One assumes that the change of state from high- to low-spin is accompanied by well-defined changes in entropy, ΔS, and enthalpy, ΔH; then the change in Gibbs energy, ΔG at temperature T is, in the usual way,

$$\Delta G = \Delta H - T\Delta S.$$

One further supposes that the ratio of the activity coefficients for the two species is unity and hence K (activities) = K (concentrations) = K.

Then

$$\Delta G = -RT \ln K$$

and

$$\ln K = \frac{\Delta S}{R} - \frac{\Delta H}{RT}.$$

Thus, if measurements of α yield a linear relationship between $\ln K$ and $1/T$, it is reasonable to assume the presence of a thermal mixture. There is substantial literature on such thermal mixtures of spin states which the reader will find referenced in the review by Iizuka and Yonetani (1970) of phenomenological aspects of spin changes in hemeproteins.

It is convenient to introduce the statistical weights (thermodynamic probabilities) W_L and W_H of the two components, ($\Delta S = R \ln W_L/W_H$), and then to define γ, the ratio of statistical weights exclusive of spin-degeneracy:

$$\frac{W_H}{W_L} = \frac{2S_H + 1}{2S_L + 1}\gamma = 3\gamma$$

where S_H and S_L are the spin quantum numbers. It follows that

$$K = \frac{\alpha}{1 - \alpha} = (3\gamma)^{-1} e^{-\Delta H/RT} .$$

A priori, one might expect γ to be of the order of unity, corresponding to some small difference in entropy. For the mixture property to be observable, significant amounts of both components must be present, or $\alpha \sim 1 - \alpha$, $K \sim 1$, and then, with $\gamma \sim 1$, $-\Delta H/RT \sim 1$ and $|\delta K/K| \sim |\delta T/T|$. Under these conditions, in the neighborhood of room temperature K would change by a percent or less per degree and the transition from one spin state to the other would take place slowly over a large range in temperature. In the latter case, unless the measurements can be made with great accuracy, observations restricted to the limited changes which occur in the temperature range available above freezing and below denaturation are unlikely to provide a certain description of the thermal behavior of the spin state distribution. It is a curious feature of those hemeprotein complexes exhibiting both spin state equilibria and continuous behavior in going through the freezing point, and for which the paramagnetic susceptibility has been measured from room down to cryogenic temperatures, that γ is not necessarily of order 1; rather $0<|\log\gamma|<7$ (Table 4-1) (Iizuka and Kotani, 1969). Apparently, there can be great differences among the numbers of sub-states belonging to the spin states of hemeproteins.

Another interesting result of the analysis of thermal equilibria between spin states is the existence of a linear relation between $\Delta S (= -R \ln 3\gamma)$ and ΔH for the various complexes of the same hemeprotein, such that the effects of the changes in entropy and enthalpy from complex to complex tend to compensate one another as in Figure 4-4. To the extent that the function (ΔS, ΔH) can be represented as a straight line through the origin, the single parameter $T_c \equiv d(\Delta H)/d(\Delta S)$, called the compensation temperature, serves to characterize all the complexes of a given hemeprotein in the following related ways:

1. It is a measure of the extent of enthalpy-entropy compensation.

2. It is the temperature at which the Gibbs energy change vanishes (i.e. $\Delta G = \Delta H - T_c \Delta S = 0$).

3. It is the temperature at which the concentrations of low- and high-spin species are equal ($\alpha = 1/2$). (This and the preceding characterization of T_c do not necessarily hold for complexes,

Table 4-1. Various parameters of the transition of spin states for ferrimyoglobin, ferrihemoglobin and ferricytochrome c peroxidase. (From Iizuka and Kotani, 1969, modified)

Complex (all Fe(III))	Ground state	γ	ΔH $(kJ \cdot mol^{-1})^a$	ΔS $(J \cdot mol^{-1} \cdot K^{-1})^a$	T_c (K)
$Mb \cdot H_2O$	High	3.05×10^{-3}	+15.23	+39.3	388
$Mb \cdot OCN^-$	High	8.53×10^{-4}	+12.69	+49.8	255
$Mb \cdot N_3^-$ (crystal)	Low	5.33×10^{1}	-15.31	-42.7	359
$Mb \cdot N_3^-$	Low	4.01×10^{1}	-15.67	-40.1	391
$Mb \cdot Im$	Low	4.12×10^{6}	-45.27	-136.8	331
$Hb \cdot OCN^-$	Low	0.328	- 2.98	+ 0.059	-
$Hb \cdot OH^-$	Low	4.20	- 7.89	-21.2	372
$Hb \cdot N_3^-$	Low	3.02×10^{2}	-21.31	-57.0	374
$CCP \cdot H_2O$	High	5.59×10^{-4}	+14.69	+53.6	274
$CCP \cdot OH^-$	Low	2.58×10^{4}	-21.84	-94.1	232

[a] 1 cal = 4.184 J

like ferric hemoglobin cyanate, for which the ΔS, ΔH couple falls close to the origin.)

When the protoheme group of myoglobin, hemoglobin, and cytochrome c peroxidase is replaced by mesoheme, T_c increases by 50 to 100° (Yonetani et al., 1972).

It can be shown that, as $|\log \gamma|$ increases, the transition in magnetic susceptibility becomes a steeper function of temperature. Thus studies of the thermodynamic properties of hemeprotein complexes for which 273 K < T_c < denaturation temperature and $|\log \gamma| \gg 1$ could be successfully carried out entirely with solutions above the freezing point.

Intermediate values for the magnetic susceptibility do not necessarily imply a thermal mixture of high- and low-spin forms. The equilibrium hypothesis must be checked by studying the temperature dependence of the susceptibility and the EPR spectra. For example, the ferricytochromes c' at physiological pH and temperature have magnetic susceptibilities significantly lower than expected for high-spin states (Ehrenberg and Kamen, 1965); the same is true at 150 and 4.2 K (Tasaki et al., 1967); the EPR spectra at 100 K suggest that the usual components of a thermal mixture are not present (Ehrenberg and Kamen 1965); the susceptibility and EPR at very low temperatures are inconsistent with the system having settled into either a high- or low-spin state

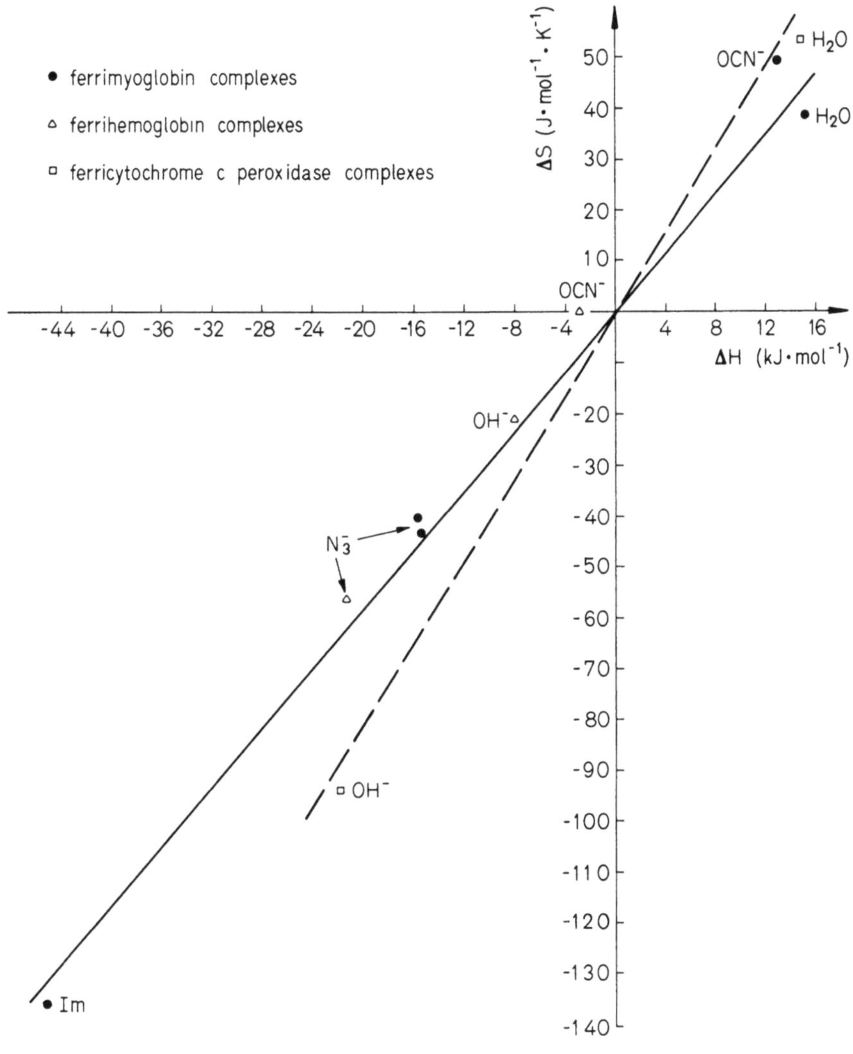

Fig. 4-4. Compensation phenomena between ΔH and ΔS. (After Iizuka and Kotani, 1969, modified)

(Maltempo et al., 1974). Maltempo (1974) proposes that, in the absence of spin-orbit coupling, the high (5/2)- and intermediate (3/2)-spin states in the ferricytochromes c' are separated by an energy Δ which is only several hundred wave-numbers, and hence that spin-orbit coupling is effective in mixing these states. Four of the five levels which result are spin-mixed doublets where the coefficients of mixing, energies, and interactions with a magnetic field depend upon the ratio of the crystal field parameter Δ to the spin-orbit coupling constant λ. The small values of Δ required for the ratio Δ/λ to generate the observed unusual magnetic properties have been explained by Maltempo (1974) in terms of the ferric ion being constrained to intermediate out-of-plane positions. Maltempo and Moss (1976) review the spin 3/2 state and quantum spin mixtures in heme proteins.

6. Influences of Symmetry upon the Energy Levels of Low- and High-Spin States

Let us consider that the metal ion is described initially by states (space plus spin) appropriate to a strong (the low-spin case) or weak (the high-spin case) field of octahedral symmetry, and take as a perturbation (in the absence both of spin-orbit coupling and an applied magnetic field) the general second order expression

$$H' = a V_z^2 + b V_x^2 + c V_y^2,$$

where $\underline{V}$ is a vector operator (either the spin $\underline{S}$ or position $\underline{r}$ for present purposes), and the axes are chosen to eliminate cross terms and labeled such that $|a| \geqq |c| \geqq |b|$. (These axes will also be the principal directions of the g-tensor, except that x and y may be interchanged.) We require that the perturbation does not change the average energy of the levels; that is, the trace of H' over the appropriate ground manifold must vanish (e.g., $\sum_{M_s=-S}^{+S} \langle M_s | H'(S) | M_s \rangle = 0$). It can be shown, as further below, that an equivalent requirement is $a + b + c = 0$. In order to arrive at the customary form for expressing the axial and rhombic components of the perturbation, we rewrite H' as follows (Bowers and Owen, 1955):

$$\begin{aligned} H' &= a V_z^2 + \frac{1}{2}(b+c)(V_x^2 + V_y^2) + \frac{1}{2}(b-c)(V_x^2 - V_y^2) \\ &= a V_z^2 + \frac{1}{2}(b+c)(V^2 - V_z^2) + \frac{1}{2}(b-c)(V_x^2 - V_y^2) \\ &= D\left(V_z^2 - \frac{V^2}{3}\right) + E(V_x^2 - V_y^2), \end{aligned}$$

where $D \equiv \frac{3}{2}a$ and $E \equiv \frac{1}{2}(a + 2b)$. The ratio of rhombic to axial strength, E/D, can be seen to have an upper limit by expressing it in terms of a and b:

$$\frac{E}{D} = \frac{1}{3} + \frac{2}{3}\frac{b}{a}.$$

But a and b are of opposite sign (unless $b = 0$), and $0 \leqq |b| \leqq \frac{1}{2}|a|$. Hence

$$0 \leqq \left|\frac{E}{D}\right| \leqq \frac{1}{3},$$

where the absolute value of E/D is taken to allow for the possibility of interchanging the x and y axes. A component ratio which exceeds the value one-third does not express the relation between rhombic and tetragonal character. If one wishes to evaluate the extent of rhombicity in such a case, it is necessary to define properly the symmetry axes.

In preparation for treating the low-spin case, we take the perturbation to be the potential

$$H' = az^2 + bx^2 + cy^2 = D(z^2 - \frac{r^2}{3}) + E(x^2 - y^2),$$

and the ground manifold to be the real, one-electron, metal orbitals d_{xy}, d_{xz} and d_{yz} (which are degenerate in an octahedral field). The diagonal matrix elements are

$$E_{xy} = \langle xy|H'|xy\rangle = D[A - \frac{1}{3}(B+B+A)] + E[B-B] = \frac{2}{3}D(A-B)$$

$$E_{yz} = \langle yz|H'|yz\rangle = D[B - \frac{1}{3}(A+B+B)] + E[A-B] = (\frac{1}{3}D - E)(B-A)$$

$$E_{xz} = \langle xz|H'|xz\rangle = D[B - \frac{1}{3}(B+A+B)] + E[B-A] = (\frac{1}{3}D + E)(B-A),$$

where the integrands yielding A are of the form $x^2\, y^2\, z^2$ and the integrands yielding B are of the form $x^4\, y^2$. Note that the trace equals zero. The "tetragonal splitting" is

$$\Delta \equiv \frac{1}{2}(E_{xz} + E_{yz}) - E_{xy} = D(B-A),$$

and the "rhombic splitting" is

$$V \equiv E_{xz} - E_{yz} = 2E(B-A).$$

Since $V/\Delta = 2E/D$, V/Δ is a measure of rhombicity only for $0 \leqq V/\Delta \leqq 2/3$. It is instructive to relate these splittings to the coefficients of the expansion of the crystal field potential in spherical harmonics,

$$H' = A_2{}^0 Y_2{}^0 + A_2{}^2(Y_2{}^2 + Y_2{}^{-2})$$

where $Y_2{}^0 = k(2z^2 - x^2 - y^2)$, $Y_2{}^2 + Y_2{}^{-2} = \sqrt{6}\,k\,(x^2 - y^2)$, and $k \equiv \sqrt{5/16\pi}$. Hence $D = 3kA_2{}^0$, $E = \sqrt{6}\,kA_2{}^2$, and $A_2{}^2/A_2{}^0 = \sqrt{3/2}\;E/D = \sqrt{3/8}\;V/\Delta$.

Turning now to the low-spin case, we note first that the ground doublet is separated from the excited sextet and quartet states by ligand field energies which are large compared with the spin-orbit coupling (λ = 420 cm^{-1} for the free ferric ion). One then starts a description of this case by excluding all states except those belonging to the ground $(t_{2g})^5$ configuration. The electron and hole level diagrams shown in Figure 4-5 presuppose that the axial component of the field is at right-angles to the heme plane and that the d_{xy} electron level is lowest, an assignment first made in the theoretical analysis by Griffith (1957) of the data from a single crystal EPR orientation study of ferrihemoglobin azide (Gibson and Ingram, 1957). The pertinent experimental feature here, one which has since been found in single crystal EPR measurements on many other low-spin hemeprotein complexes (Helcké et al., 1968; Hori, 1971; Mailer and Taylor, 1972),

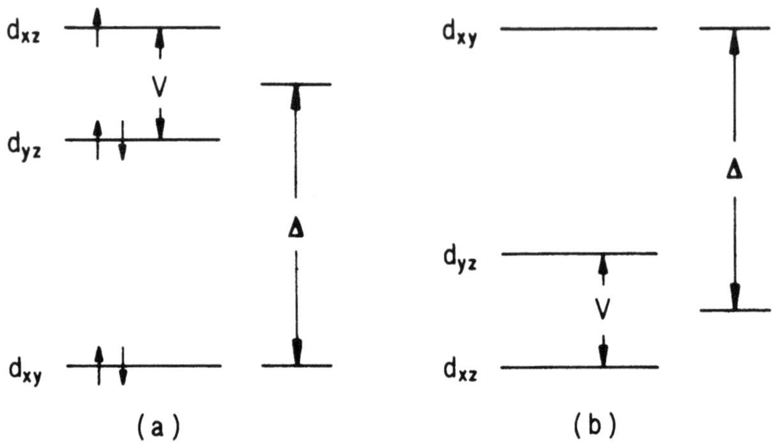

Fig. 4-5a and b. Low-spin ferric ion energy levels. (a) Single electron levels, showing occupancies. (b) Hole levels

is that the direction of the maximum g-value lies along or within several degrees of the heme normal. In ferrimyoglobin complexes the direction of g_{min} does not appear to be determined by the projection of the imidazole ring (in coordination position 5) upon the heme plane (Hori, 1971), as was early suggested. Since Δ, V, and λ can all be of the same order of magnitude, the perturbation of the three-fold degenerate ground manifold is taken to include the spin-orbit and ligand field energies simultaneously. Three Kramers doublets and their energies are obtained (in terms of an effective λ) from diagonalization of the resulting matrix, and g-values are calculated in the usual way for the lowest Kramers doublet (which, in general, is separated from the other two doublets by energies very large compared with the Zeeman energy). From this simple model, which does not explicitly take into account molecular orbital formation or the influence of the excited states of high and intermediate spin (Griffith, 1971), g-values can often be calculated which are close to those observed experimentally (Loew, 1970; Weissbluth, 1974). The ratios V/Δ and λ/Δ which produce the desired g-values then provide some basis for a discussion of structural differences among the low-spin complexes studied. In spite of the great advance in perspective which attended the introduction of molecular orbitals into the treatment of magnetic properties of "simple" covalent octahedral complexes (Stevens, 1953), there have been only a few attempts to treat realistically the low-spin ferric ion in hemeproteins (Horrocks and Greenberg, 1973; Shulman et al., 1971).

The zero-field spin-Hamiltonian for the high-spin case

$$D[S_z^2 - \frac{1}{3}S(S+1)] + E(S_x^2 - S_y^2)$$

will be discussed in Chapter 6-6; it has zero trace and rhombic to tetragonal ratio E/D. Because E/D is small (< .06) in hemeproteins, there are well-defined g-values for magnetic field strengths in the range customarily employed for EPR spectroscopy. (Blumberg (1967) discusses the EPR of the high-spin ferric ion in fields of arbitrary rhombicity.) The g-tensors for ferric myoglobin and hemoglobin (acid forms) are very close to axial, a curious feature which they do not share with other high-spin ferric hemeproteins or with some abnormal hemoglobins (Bemski and Nagel, 1968; Ehrenberg, 1962; Ehrenberg and Estabrook, 1966; Hayashi et al., 1966; Morita and Mason, 1965; Peisach et al., 1971). These differences in E/D have not yet been assigned to specific structural differences. The zero-field splittings of the high-spin ground sextet are in the range 10 to 100 cm^{-1} while the ligand field splittings of the t_{2g} levels in the low-spin complexes are in the range 400 to 2000 cm^{-1}. Kotani (1968) discusses the theory of the zero-field splittings for both the high- and low-spin cases.

7. Ligand Hyperfine Effects in Ferric Hemeproteins

The interactions of the unpaired spins of the metal ion with the nuclear magnetic dipoles and electric quadrupoles of nearby atoms provide a rich source of structural information. In the low-spin case, most of the data obtained thus far has come from NMR experiments (Wüthrich, 1970), and some from ENDOR (Scholes and van Camp, 1976). EPR and ENDOR measurements have been successful in quantitating ligand hyperfine interactions in high-spin hemeproteins, and these studies form the basis for the discussion of principles to be given in this section.

The major interactions between the unpaired electrons and the four pyrrole nitrogens can be expressed in terms of the ligand hyperfine Hamiltonian:

$$H = \sum_{i=1}^{4}\Big(H^{i}_{\text{Fermi}} + H^{i}_{\text{covalent dipole-dipole}} + H^{i}_{\text{ionic dipole-dipole}} + H^{i}_{\text{quadrupole}}\Big).$$

Some unpaired spin from the ferric d-shell is delocalized onto the nitrogens, giving rise through the s-character of the atomic hybrid orbitals to Fermi contact or "isotropic" ligand hyperfine interactions, H^{i}_{Fermi}, and through p-character to "covalent" dipolar (including orbital field) anisotropic interactions, $H^{i}_{\text{covalent dipole-dipole}}$. Most of the unpaired spin remains on the metal and, from this somewhat distant volume, interacts with the nitrogen nuclear magnetic moments to produce dipolar energies $H^{i}_{\text{ionic dipole-dipole}}$. The energy of the system depends also upon the nitrogen nuclear spin states through the quadrupole moments of these nuclei; this interaction of the quadrupole moment with the electric field gradient produced at the nucleus by asymmetric charge distribution in the ligand valence shell is represented above by $H^{i}_{\text{quadrupole}}$.

The Fermi interaction is sufficiently larger than the other three that the system can be initially characterized by diagonalization of the four contact terms

$$H^{i}_{\mathrm{Fermi}} = A^{i}_{F}\,\underline{I}^{N}_{i} \cdot \underline{S}\,, \quad i = 1-4.$$

We prefer here to employ ground electronic spin states for rhombic symmetry rather than the simpler ones for axial symmetry derived in Chapter 6-6. The rhombic functions are

$$|+\rangle = \frac{1}{\sqrt{2g}}\left[\frac{\sin\theta\,(g_x\cos\phi - ig_y\sin\phi)}{\sqrt{g-g_z\cos\theta}}\,|\tfrac{5}{2},\tfrac{1}{2}\rangle + \sqrt{g-g_z\cos\theta}\,|\tfrac{5}{2},-\tfrac{1}{2}\rangle\right]$$

$$|-\rangle = \frac{1}{\sqrt{2g}}\left[\frac{\sin\theta\,(ig_y\sin\phi - g_x\cos\phi)}{\sqrt{g+g_z\cos\theta}}\,|\tfrac{5}{2},\tfrac{1}{2}\rangle + \sqrt{g+g_z\cos\theta}\,|\tfrac{5}{2},-\tfrac{1}{2}\rangle\right],$$

where the direction of the applied magnetic field is given by the polar angles θ and ϕ in the g_x, g_y, g_z system and where small admixtures (from a rhombic term in the spin-Hamiltonian, if present, spin-orbit coupling with excited states, and from the applied magnetic field when $\theta \neq 0$) of states other than $|\frac{5}{2},\frac{1}{2}\rangle$ and $|\frac{5}{2},-\frac{1}{2}\rangle$ have been neglected. The angular dependence of g is

$$g^2 = g_z^2\cos^2\theta + g_x^2\sin^2\theta\cos^2\phi + g_y^2\sin^2\theta\sin^2\phi.$$

Table 4-2 gives the matrix elements of $\underline{S}$. The spin of the nitrogen nucleus being 1, for each nitrogen i there is a manifold of three eigenstates of the Fermi interaction, say p_i, q_i, r_i where p, q, and r correspond to projection quantum numbers (1, 0, -1). For the $|-\rangle$ electronic spin state these eigenvectors are, with energies W,

$$|q\rangle = \frac{\sqrt{2}}{G}[-3\sin\theta\,(g_x\cos\phi - ig_y\sin\phi)\,|1\rangle + \sqrt{2}g_z\cos\theta\,|0\rangle$$

$$+ 3\sin\theta\,(g_x\cos\phi + ig_y\sin\phi)\,|-1\rangle],$$

$$W_q = 0;$$

$$|p\rangle = \frac{1}{2G}\left[\frac{(G-2g_z\cos\theta)}{g_\perp}(g_x\cos\phi - ig_y\sin\phi)\,|1\rangle - 6\sqrt{2}g_\perp\sin\theta\,|0\rangle\right.$$

$$\left. + \frac{(G+2g_z\cos\theta)}{g_\perp}(g_x\cos\phi + ig_y\sin\phi)\,|-1\rangle\right],$$

$$W_p = \frac{A_F G}{4g};$$

$$|r\rangle = \frac{1}{2G}\left[\frac{(G + 2g_z\cos\theta)}{g_\perp}(g_x\cos\phi - ig_y\sin\phi)\,|1\rangle + 6\sqrt{2}g_\perp\sin\theta\,|0\rangle + \frac{(G - 2g_z\cos\theta)}{g_\perp}(g_x\cos\phi + ig_y\sin\phi)\,|-1\rangle\right],$$

$$W_r = -\frac{A_F G}{4g}$$

where $g_\perp^2 \equiv g_x^2\cos^2\phi + g_y^2\sin^2\phi$
and $G^2 = 4g_z^2\cos^2\theta + 36\sin^2\theta(g_x^2\cos^2\phi + g_y^2\sin^2\phi)$. (The exact values of W_p and W_r are a few per cent larger than those calculated with G and a few per cent smaller than those calculated with the G_{eff} of Scholes (1970), equation A7.) Table 4-3 gives the matrix

Table 4-2. Matrix elements of S

	+	−
+	$\frac{3g_x\sin\theta\cos\phi}{2g}$	$\frac{3}{2g}\left(\frac{g_z g_x\cos\theta\cos\phi + ig\,g_y\sin\phi}{g_\perp}\right)$
−	$\frac{3}{2g}\left(\frac{g_z g_x\cos\theta\cos\phi - ig\,g_y\sin\phi}{g_\perp}\right)$	$\frac{-3g_x\sin\theta\cos\phi}{2g}$

S_x

	+	−
+	$\frac{3g_y\sin\theta\sin\phi}{2g}$	$\frac{3}{2g}\left(\frac{g_z g_y\cos\theta\sin\phi - ig\,g_x\cos\phi}{g_\perp}\right)$
−	$\frac{3}{2g}\left(\frac{g_z g_y\cos\theta\sin\phi + ig\,g_x\cos\phi}{g_\perp}\right)$	$\frac{-3g_y\sin\theta\sin\phi}{2g}$

S_y

	+	−
+	$\frac{g_z\cos\theta}{2g}$	$\frac{-g_\perp\sin\theta}{2g}$
−	$\frac{-g_\perp\sin\theta}{2g}$	$\frac{-g_z\cos\theta}{2g}$

S_z

Table 4-3. Matrix elements of I

	p	q	r
p	$-6\sin\theta g_x\cos\phi$	$\frac{1}{\sqrt{2}}\frac{1}{g_\perp}(2g_xg_z\cos\theta\cos\phi-iGg_y\sin\phi)$	0
q	$\frac{1}{\sqrt{2}}\frac{1}{g_\perp}(2g_xg_z\cos\theta\cos\phi+iGg_y\sin\phi)$	0	$\frac{1}{\sqrt{2}}\frac{1}{g_\perp}(2g_xg_z\cos\theta\cos\phi-iGg_y\sin\phi)$
r	0	$\frac{1}{\sqrt{2}}\frac{1}{g_\perp}(2g_xg_z\cos\theta\cos\phi+iGg_y\sin\phi)$	$6\sin\theta g_x\cos\phi$

$G\cdot I_x$

	p	q	r
p	$-6\sin\theta g_y\sin\phi$	$\frac{1}{\sqrt{2}}\frac{1}{g_\perp}(2g_yg_z\cos\theta\sin\phi+iGg_x\cos\phi)$	0
q	$\frac{1}{\sqrt{2}}\frac{1}{g_\perp}(2g_yg_z\cos\theta\sin\phi-iGg_x\cos\phi)$	0	$\frac{1}{\sqrt{2}}\frac{1}{g_\perp}(2g_yg_z\cos\theta\sin\phi+iGg_x\cos\phi)$
r	0	$\frac{1}{\sqrt{2}}\frac{1}{g_\perp}(2g_yg_z\cos\theta\sin\phi-iGg_x\cos\phi)$	$6\sin\theta g_y\sin\phi$

$G\cdot I_y$

	p	q	r
p	$-2g_z\cos\theta$	$-\frac{6}{\sqrt{2}}g_\perp\sin\theta$	0
q	$-\frac{6}{\sqrt{2}}g_\perp\sin\theta$	0	$-\frac{6}{\sqrt{2}}g_\perp\sin\theta$
r	0	$-\frac{6}{\sqrt{2}}g_\perp\sin\theta$	$2g_z\cos\theta$

$G\cdot I_z$

elements of $\underline{I}$. The tetrapyrrole nitrogen nuclear spin state is specified by four of the individual states, e.g. (q_1, p_2, q_3, r_4) with total ligand projection quantum number M = O. It is convenient to designate such a state by $|{}^{qp}_{rq}\rangle$ where the square array of nuclear state symbols indicates the array of nitrogens in the heme group. Differences among the Fermi coefficients A_F^i are likely to arise from small differences among the delocalizations onto the nitrogens and possibly some differences in hybridization among the nitrogens as well. In such case, the contact energies of, for example, $|{}^{qp}_{rq}\rangle$ and $|{}^{pr}_{qq}\rangle$, which have the same M, would not be exactly the same. In ferrimyoglobin, the resolution of g = 2 ENDOR lines of the heme nitrogens suggests that these differences do not exceed a few per cent (Scholes et al., 1972). Larger energy differences arise from the quadrupole interactions. If one takes a principal axis of the quadrupole tensor to parallel the heme normal, then $H^i_{\text{quadrupole}} = Q^i[(I_{i_{xo}}^N)^2 - \frac{1}{3}I_i^N(I_i^N+1) + \eta\{(I_{i_{yo}}^N)^2 - (I_{i_z}^N)^2\}]$ where xo lies along the $N_{i+2} - N_i$ direction, yo along $N_{i+3} - N_{i+1}$, and η is an asymmetry parameter. When $H^i_{\text{quadrupole}}$ is averaged over the eigenstates of H^i_{Fermi}, the energies of p and r are changed equally and by a different amount than q, but these energy changes are the same for both $|+\rangle$ and $|-\rangle$ and hence do not affect the line position. The next order perturbation calculation with $H^i_{\text{quadrupole}}$ does result in a spectral shift of p and r from q. Thus $|{}^{qq}_{qq}\rangle$ and $|{}^{qp}_{rq}\rangle$, which have the same M, have electronic Zeeman levels of slightly different separation. The anisotropic dipole-dipole interactions also produce spectral shifts among states of the same M (Scholes, 1970). It can be shown that these shifts vanish in the principal directions. In those hemeproteins where there is an imidazole nitrogen in position 5 of the coordination sphere, the bonding is not equivalent and the energies of interaction of its nucleus with the metal ion are independent and different from those of the porphyrin nitrogens. (Other nuclear dipoles, in nearby water molecules and amino acid residues, and ferric dipoles, in neighboring hemes, also produce energy differences which contribute to line broadening.)

Let us imagine a high-spin heme group free of ligands in positions 5 and 6, and consider only the Fermi interactions with all four A_F^i the same. For a given electronic state, $|+\rangle$ or $|-\rangle$, there are 19 states of hyperfine energy W_q (= O) corresponding to M = O, sixteen each of energies W_p and W_r (M = ±1), ten each of energies $2W_p$ and $2W_r$ (M = ±2), four each of energies $4W_p$ and $4W_r$. If resolved in the EPR spectrum, there would then be a set of nine hyperfine lines, weighted 1:4:10:16:19:16:10:4:1. Thus far this ligand hyperfine pattern has been seen only in the EPR spectra from ferriprotoporphyrin IX dimethyl ester chloride and hydroxide doped into single crystals of the aromatic hydrocarbon perylene (Scholes, 1970) (Fig. 4-6). The hyperfine splittings would be expected to exhibit the angular variation

$$\Delta H = \frac{\Delta E}{g\beta} = \frac{2W}{g\beta} = \frac{A_F}{2\beta}\,\frac{G(\theta,\phi)}{g^2(\theta,\phi)},$$

a function which goes through a sharp maximum at $\theta \approx 20^\circ$, and the data does show this behavior (Fig. 5 of Scholes, 1970). When

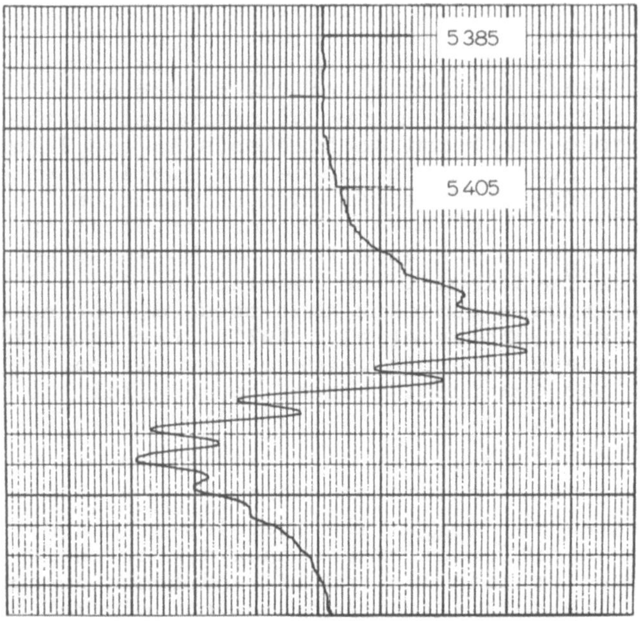

Fig. 4-6. EPR first-derivative spectrum for hemin in perylene. 4.2 K. 34.40 GHz. Numbers indicate magnetic field strength in Gauss. Splitting of hyperfine lines is about 4 Gauss. (From Scholes, 1970)

$H^{i}_{\text{ionic dipole-dipole}}$ is considered, an additional parameter is introduced, and the observed angular variation can be fit almost exactly. The value $2.9 \times 10^{-4}\text{cm}^{-1}$ for A_F is obtained from these measurements, for both the chloride complex and the hydroxide, to which corresponds 2.7% of an unpaired spin in a nitrogen 2s orbital and 8% in an sp^2 trigonal hybrid. Thus, about 32% of an unpaired spin is delocalized from the ferric ion onto the four porphyrin nitrogens. This number can be compared with the delocalization from the cupric ion in protoporphyrin IX dimethyl ester (Table 3-3); the two differ by less than 10%. For both metal ions, the molecular orbital is the σ-bonding one involving mainly $d_{x^2-y^2}$.

While ligand hyperfine structure has not yet been resolved in EPR spectra from hemeproteins, successful ENDOR measurements have been made on a frozen solution of ferrimyoglobin (Scholes et al., 1972) and it is useful to compare the results with those above from the EPR of heme oriented in an organic crystalline environment. In Figure 4-7 are shown ENDOR spectra from hemin chloride in dimethyl formamide-methanol (1:1, v/v) and from ferrimyoglobin in 50% (v/v) glycerol, 0.1 M phosphate buffer, pH 6. The ENDOR experiment is done with the magnetic field fixed at an EPR resonance which is exhibiting saturation under constant microwave power. The frequency of a radio frequency field applied simultaneously to the sample is swept through the region where, in this case, nitrogen nuclear spin transitions occur, and the amplitude of the EPR signal is monitored. (The ENDOR signal is caused by the radio frequency field but is observed as an increase in the microwave absorption. The latter change, in essence, is due to an effective decrease in spin-lattice relaxation time which arises from the increased availability of alter-

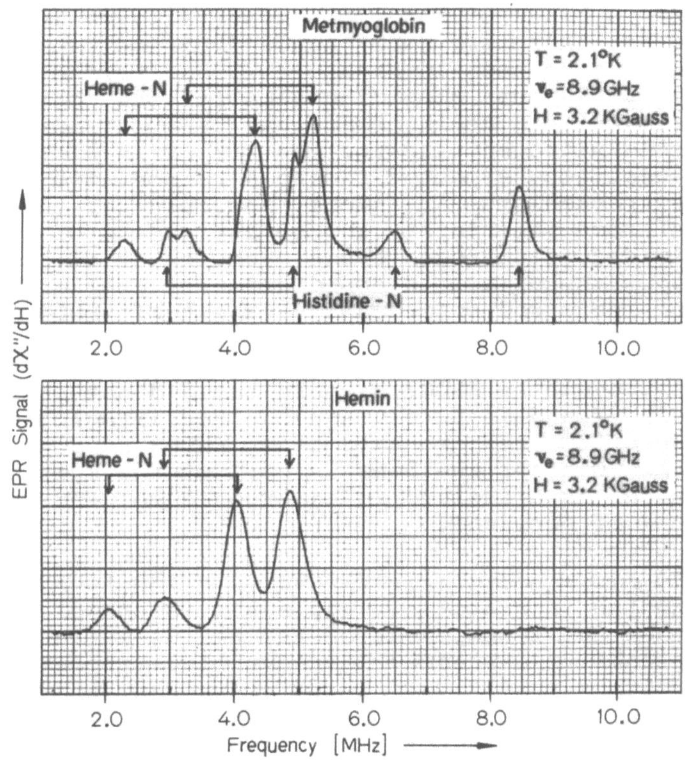

Fig. 4-7. Part of the ENDOR spectrum of ferrimyoglobin and hemin. The absence of the histidine nitrogen in hemin reduces the eight-line spectrum of ferrimyoglobin to four lines. The assignment of the lower frequency pair of the imidazole is tentative. (From Scholes et al., 1972)

nate paths to the lower electronic spin state.) In the experiments of Figure 4-7, the magnetic field strength was set near the g = 2 resonance (high-field extremum), where the EPR signal comes from only those molecules with heme normals parallel to the field, a situation like that which can be produced in measurements from a single crystal. The spin-Hamiltonian for one of the coordinated nitrogens can then be written

$$H = A_{zz}I_zS_z + Q_{zz}[I_z^2 - \frac{1}{3}I(I+1)] - g_N\beta_n HI_z,$$

where the first term includes $H^i_{\text{Fermi}} + H^i_{\text{covalent dipole-dipole}}$ $(+ H^i_{\text{ionic dipole-dipole}})$, the second term is $H^i_{\text{quadrupole}}$ with the non-axial part omitted (justified by the data), the last term is the nuclear Zeeman energy with g_N the nuclear g-value for nitrogen, and z is the axis normal to the heme plane. Diagonalization of this Hamiltonian gives the ENDOR transition frequencies, ν_{ENDOR} $(I = 1, \Delta M_I = +1)$,

$$h\nu_{\text{ENDOR}} = \frac{1}{2}A_{zz} \pm P_{zz} \pm g_N\beta_n H,$$

of which there are four occurring in two pairs. Since $g_N\beta_n/h$ is known for ^{14}N(0.308MHzG^{-1}), the members of the pairs are readily identified. In the case of ferrimyoglobin, in addition to the

ENDOR bands of the four essentially equivalent porphyrin nitrogens, there are ENDOR bands from the bound imidazole nitrogen, as can be seen in Figure 4-7. Analysis of this data shows that $|A_{zz}|$ is just about the same for the porphyrin nitrogens in ferrimyoglobin as in hemin, but is about 50% greater for the imidazole nitrogen than for the porphyrin nitrogens. Because of the anisotropy expected in the A-tensor, the Fermi part cannot be extracted from A_{zz} alone. Possibly for this reason, or because of some uncertainty in the assignment of the two lower frequency imidazole nitrogen bands, Scholes et al. (1972) do not comment upon the direct interpretation of the larger $|A_{zz}|$ value in terms of increased spin density. The hemin ENDOR is somewhat solvent dependent, $|A_{zz}|$ averaging close to the value found in perylene crystals. The presence of 50% glycerol in the protein solution improved the ENDOR signals but did not change the ENDOR frequencies found in purely aqueous ferrimyoglobin solution. With respect to the heme nitrogens, ENDOR at g = 2 shows them to be the same whether the heme is part of the myoglobin molecule in solution, is dissolved in an organic solvent, or dispersed in a crystal of perylene.

$H^{i}_{\text{ionic dipole-dipole}}$ depends upon the radius vector from nucleus i to the iron atom. If nucleus i is one of the pyrrole nitrogens, then the z-component of the radius vector is the iron out-of-planarity distance. Tetrapyrrole nitrogen nuclear spin states are not pure; $H^{i}_{\text{ionic dipole-dipole}}$ and $H^{i}_{\text{quadrupole}}$ connect any one such state to a number of others. Under certain conditions, the fundamental (intrinsic spin-packet) linewidth is determined by tetrapyrrole nitrogen nuclear spin state mixing, and the angular dependence of this linewidth can then be quantitatively related to iron out-of-planarity (Brill, 1972). There is the possibility that the angular dependence of power saturation and spin-lattice relaxation time together can be used to provide a sensitive measure of iron out-of-planarity (Brill et al., 1975).

8. Iron Hyperfine Effects

Among the isotopes of iron with significant abundance, only ^{57}Fe, 2.2% abundance, has a nonzero nuclear spin. Mössbauer spectra can be obtained from iron proteins without ^{57}Fe enrichment, although this is sometimes necessary, and used to investigate aspects of the bonding of the iron which influence the hyperfine levels. No attempt will be made here to cover recently reviewed material on the applications of Mössbauer spectroscopy to heme proteins (Lang, 1970; Weissbluth, 1974). In the case of hemoglobin, the ability to prepare stable α and β subunits has led to a new set of Mössbauer experiments. An initial finding from these investigations is that isolated deoxygenated α and β chains, which have a significantly greater oxygen affinity than intact deoxyhemoglobin-A, have essentially the same Mössbauer spectra as deoxyhemoglobin-A, both with regard to isomeric shift and quadrupole splitting, and also in the temperature dependence of the latter (Huynh et al., 1974). This result indicates that the differences in oxygen affinity of the subunits in fully deoxygenated intact hemoglobin as compared with their isolated forms are not due to differences in the electronic states of the ferrous ions. In Mössbauer spectra from oxygenated hemoglobin, the quadrupole moment can be seen to decrease with increasing temperature. Mössbauer measurements on certain model heme compounds that can be oxygenated exhibit this same behavior (Collman et al., 1975). The three-dimensional structure of these compounds is available (Collman et al., 1974) and shows that the oxygen molecule is bound in two distinct orientations. The temperature dependence of the Mössbauer spectrum can be interpreted in terms of the oxygen molecule jumping between the two states.

9. Modified Hemes

For those hemeproteins from which the heme group can be removed and replaced without changing the structure or functional ability (e.g. hemoglobin, myoglobin, horseradish peroxidase), there is the possibility of substituting modified hemes for the original and then investigating the effects of these substitutions upon structural and functional properties. A minimal modification is the one mentioned in the last section, ^{57}Fe for ^{56}Fe, carried out to improve the signal-to-noise ratio of Mössbauer measurements. A major modification is the substitution of cobalt for iron. Cobalt hemoglobin and myoglobin bind oxygen (Hoffman and Petering, 1970). Yamamoto et al. (1976) prepared hybrid hemoglobin containing iron and cobalt protoporphyrin IX in α- and β-subunits, $\alpha(Fe)_2\beta(Co)_2$, and its complement, $\alpha(Co)_2\beta(Fe)_2$. They have obtained oxygen equilibrium curves spectrophotometrically and analyzed these to arrive at two sets (one for $\alpha(Fe)_2\beta(Co)_2$ and the other for $\alpha(Co)_2\beta(Fe)_2$) of the twelve equilibrium constants governing the distribution of oxygen among the nine possible states of the tetramers. Manganese (II) protoporphyrin IX has been recombined with apoproteins of cytochrome c peroxidase, horseradish peroxidase, myoglobin, and hemoglobin, and the spectral properties of these high-spin artificial hemeproteins measured (Yonetani et al., 1970).

The influence of porphyrin sidechains on the physical properties and biochemical activity of hemeproteins has been the subject of many studies. Structures of some of the porphyrins used are shown in Figure 4-8. First, however, it should be noted that, in addition to the iron-imidazole link and to many hydrophobic attractive interactions between porphyrin and protein, the proprionyl carboxyl groups of porphyrins are implicated in the binding of heme to protein. In the case of horseradish peroxidase, for example, the peak of the Soret band obtained when protoporphyrin IX reacts with the apo-enzyme at pH 6.8 has the same position and nearly the same absorptivity as that of the holoenzyme. If the carboxyl groups of either protohemin or protoporphyrin IX are blocked by forming the dimethyl esters, then there is no spectrophotometric indication of binding when the modified prosthetic group is added to the apoenzyme (Maehly, 1961). (It has also been shown that protoporphyrin competes with protoheme for the

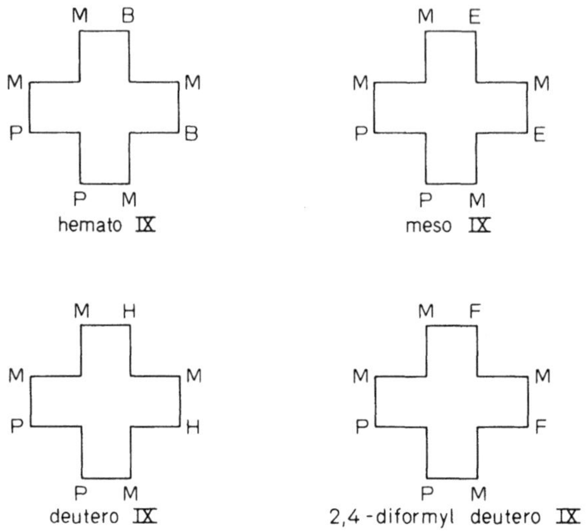

Fig. 4-8. Structures of some of the porphyrins used in producing modified hemes

M $-CH_3$
B $-CHOHCH_3$
P $-CH_2CH_2COOH$
E $-CH_2CH_3$
F $-CHO$

same sites on apo-horseradish peroxidase, and that the dimethyl ester of protoheme combines with globin.) While the ferric ion cannot be replaced by the other metal ions without loss of activity in horseradish peroxidase, the substitution of mesoheme and hematoheme for protoheme results in somewhat greater activity in catalyzing the peroxidation of pyrogallol (Paul, 1959). Spectral and functional studies of modified cytochrome c peroxidases have given similar results. In particular free propionic acid groups are necessary for the binding of iron-free protoporphyrin to the apo-enzyme; protoheme monomethyl ester does combine, but the latter complex has only 1/20 the activity of the native enzyme (Asakura and Yonetani, 1972); the substitution of hemato-, meso-, and deuterohemins for protohemin has little effect upon the activity. As mentioned in Chapter 4-5, the compensation temperature for mesohemeproteins is 50-100° higher than for the native protohemeproteins. The two isomers of monoformyl-monovinyl- and 2,4-diformyl-deuterohemins have similar absorption spectra when free in solution; but when these groups are bound to apomyoglobin and the oxygenated complex studied, spectral differences are found, and there are also differences in oxygen affinity among these synthetic formylmyoglobins (Sono and Asakura, 1975). Clearly the interactions of the porphyrin sidechains with the protein moiety can play a role both in the electronic structure and function of the heme group.

10. Photodissociation and Recombination

The classic experiments of Warburg (1949) on the identification of the "respiratory enzyme" in yeast and acetic acid bacteria were based upon systematic studies of the absorption spectra and photodissociation of the carbon monoxide complexes of hemes and hemeproteins. The respiratory enzyme (cytochrome oxidase) is inhibited by carbon monoxide, and respiration is restored by light (monochromatic light at many wavelengths is used in the experiments). Because the quantum yield is nearly constant throughout the spectral region investigated, the rate of respiration at each wavelength is proportional to the absorptivity of the carbon monoxide complex. The relation between the rate of a photo-induced biological process and light intensity as a function of wavelength is called the "action spectrum". (A general treatment of action spectra and quantum yields can be found in Setlow and Pollard, 1962.) The dissociation of the carbon monoxide complex of myoglobin has been so well characterized that it can be used as an accurate measure of light intensity (Olson and Chance, 1960).

The optical spectra of the compounds of cytochrome oxidase with oxygen have been difficult to measure because of their instability; at temperatures above 0°C the fraction of enzyme present as oxygen compounds is only of the order of 10%. Chance et al. (1975) have arrived at a technique for observing the oxygen compounds which depends upon three factors: (1) in ethylene glycol solution at -18°C the carbon monoxide compound of cytochrome oxidase is stable for several minutes against oxygen replacement; (2) the carbon monoxide compound can be dissociated by a laser flash; (3) the free enzyme reacts readily in the frozen state with oxygen dissolved in the solution before freezing. The technique consists of preparing the CO complex in the ethylene glycol at -18°C, mixing the solution with oxygen and lowering the temperature an additional 80°, dissociating the CO complex with a laser flash and observing optically the combination with oxygen and succeeding processes. At -100°C, the light-insensitive first species, which has a spectrum like oxyhemoglobin, is stable. As the temperature is raised, transfer of electrons to the oxygen occurs, (eventually resulting in the formation of water), and is accompanied by a spectral sequence which reflects valence changes in the heme and copper sites, and concomitant structural changes (of uncertain nature at the present time).

The light-induced oxidation of cytochrome in photosynthetic bacteria at low temperatures is in some ways related to the experiments just described. The rate of such oxidation has been spectrophotometrically measured following absorption of a short laser pulse (De Vault and Chance, 1966). Above 120 K, the temperature dependence follows Arrhenius behavior, but below 100 K the rate is essentially temperature-independent and slow (Fig. 4-9). Quantum-mechanical "tunneling" was suggested to explain the small and constant rate of oxidation in the low temperature range. (See also De Vault et al., 1967.)

The kinetics of the recombination of carbon monoxide and oxygen with myoglobin and related hemeproteins has been measured at temperatures in the range 2 K to 350 K (Alberding et al., 1976; Austin et al., 1975). A laser is used to flash photodissociate the complex, and the subsequent rebinding is observed optically. A transient analyzer with a logarithmic time base is used which, in a single sweep, goes from 2 μs to 1 ks. Figure 4-10 shows the

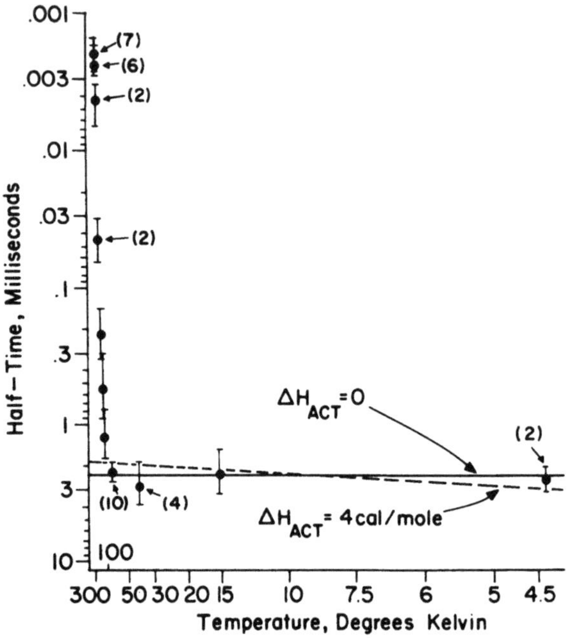

Fig. 4-9. Temperature dependence of the rate of laser-induced cytochrome oxidation in *Chromatium*. The logarithm of the reaction rate versus 1/T is plotted. Numbers in parentheses: numbers of observations averaged. (From de Vault et al., 1967)

rebinding of CO and O_2 to myoglobin in glycerol-water over this time range, as a function of temperature. N(t) is the fraction of myoglobin that is free of CO (or O_2) t seconds after the flash. Five processes can be distinguished. In only one, namely that which occurs above 210 K, is the rate dependent upon ligand (CO or O_2) concentration; this rebinding is exponential in a liquid glycerol-water solvent. All the other processes have concentration-independent rates; they are intramolecular. The one which dominates below 25 K is quantum-mechanical tunneling. Thus, in Figure 4-11(c), the rate is seen to be essentially temperature independent below 10 K. Figure 4-11(a) is a model potential barrier: below 25 K tunneling is the greater contributor to the rate constant, while above 25 K, temperature-dependent barrier jumping predominates. However, the process between 25 K and 160 K is not exponential but is given, approximately, by $N(t) = (1 + t/t_0)^{-n}$, where t_0 and n are temperature-dependent. The two other ligand-concentration independent processes, emerging at about 170 K and 200 K, are somewhat more exponential. The non-exponential behavior of the process between 25 K and 160 K can be explained on the basis of a distribution in activation energies, such as the one shown in Figure 4-11(d). N(t) is then given by

$$N(t) = \int_0^\infty dE\, g(E) \exp(-kt).$$

All told, there are four barriers which a carbon monoxide molecule encounters, one after the other, on its way in from the solvent to the ferrous ion binding site, as schematized in Figure 4-12(a). A pictorial interpretation of the structural features responsible for these barriers is shown in Figure 4-12(b)

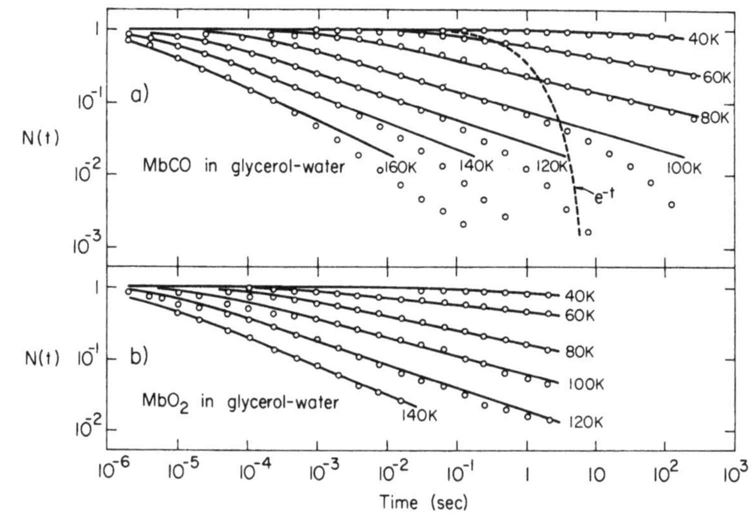

Fig. 4-10. Rebinding of CO and O_2 to myoglobin after photodissociation. (From Austin et al., 1975)

with labels A through E corresponding to the wells similarly labeled in Figure 4-12(a). Thermodynamic parameters characterizing each barrier can be estimated from the rate data and its temperature dependence. These parameters can then be related to ligand specificity and to ordering/disordering processes in the protein. For example, at 310 K the heights of the four free energy barriers for carbon monoxide progressively increase on the way in (due to strongly decreasing entropy from the top of barrier III to the bound state, well A), and carbon monoxide molecules are partially rejected at each barrier. At the same temperature oxygen molecules see mainly three barriers of equal

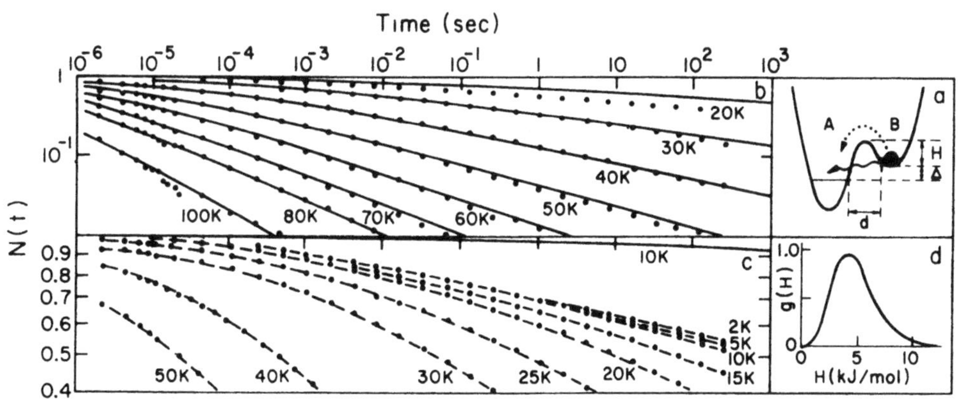

Fig. 4-11. (a) A molecule in well B can move to A by hopping over the barrier or by tunneling through it. (b) Rebinding of CO to β Hb after photodissociation. (c) As (b), but with expanded N(t) scale. The solid line labeled 10 K indicates N(t) as expected without tunneling. (d) Activation energy spectrum for β HbCO. (From Alberding et al., 1976)

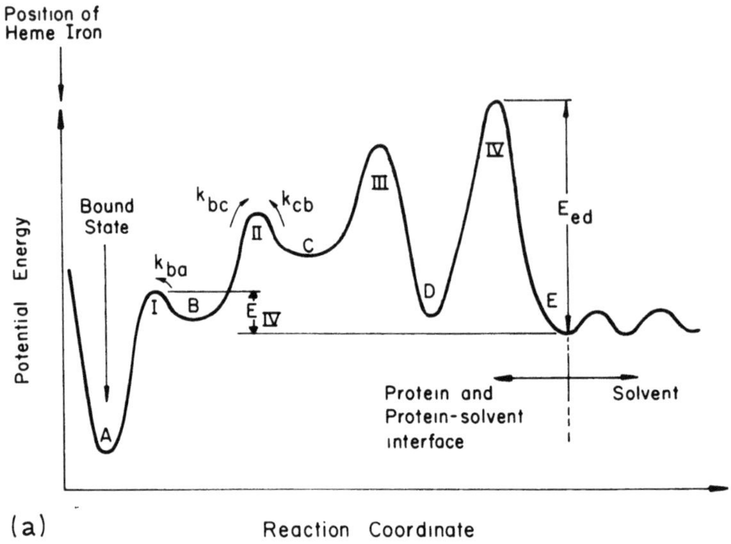

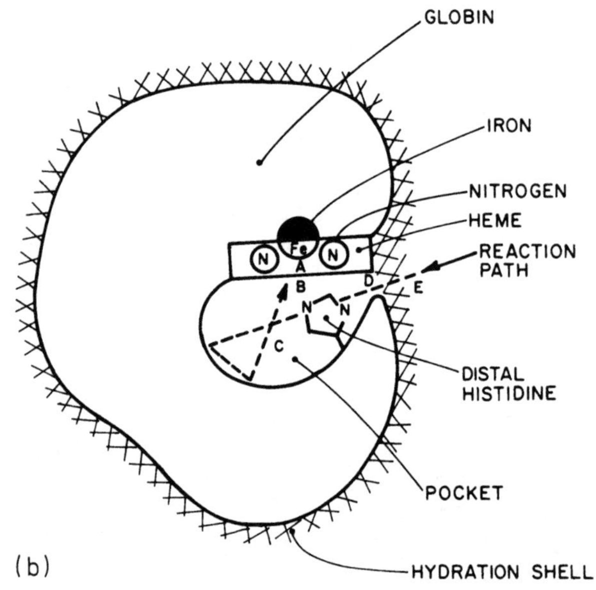

Fig. 4-12. (a) Barrier which carbon monoxide molecule encounters on way from solvent to ferrous ion binding site. (b) Pictorial representation of structural features responsible for barrier of (a). (From Austin et al., 1975)

height, and move relatively smoothly in to the iron. In related measurements, the recombination kinetics of carbon monoxide with protoheme have been measured as a function of temperature (Alberding et al., 1976). A two-barrier model, corresponding to the first and last of Figure 4-12, is consistent with the data. As with myoglobin (and the β-chain of hemoglobin), the nonexponential rebinding at low temperatures is explained in terms of distributed activation enthalpy and entropy for the inner barrier.

Chapter 5

Nonheme Iron and Molybdenum

1. Iron Storage and Transport Proteins

The electronic structure of the iron in these proteins is the subject of a recent review by Oosterhuis (1974).

2. Iron-Sulfur Proteins

Much of the literature on iron-sulfur proteins up to 1973 is summarized and referenced in a two volume treatise edited by Lovenberg (1973), of which the second volume deals with molecular properties. More compact reviews of structure and electronic states are by Bearden and Dunham (1970); Beinert (1972); Palmer (1975); Sands and Dunham (1974) and Tsibris and Woody (1970). Exchange interactions among metal ions play an important role in iron-sulfur proteins; a review of the theory of metal-metal exchange interactions can be found in Kokoszka and Gordon (1969). The topics to be treated below illustrate only a few salient features of iron-sulfur complexes.

The early magnetic and optical measurements on xanthine oxidase, a complex enzyme which contains molybdenum and flavin as well as eight iron atoms preceded any knowledge of the involvement of sulfur. Kinetic magnetic susceptibility measurements made during the reaction of the enzyme with xanthine showed an increase in paramagnetism beyond that which could be accounted for on the basis of all the prosthetic groups; the generation of substrate free radicals was therefore suggested (Ackerman and Brill, 1962). On the basis of present knowledge, discussed later in this section, a major component of the increase in paramagnetism could be the conversion upon reduction of largely diamagnetic antiferromagnetically coupled ferric ions to paramagnetic antiferromagnetically coupled ferrous-ferric clusters. Bray et al. (1959) examined the xanthine-xanthine oxidase system by EPR and found a complex sequence of changes, all in the region $1.9 < g < 2.1$. Much of the interpretation of the xanthine oxidase data involves molybdenum, to be discussed in Chapter 5-3.

Beinert and Sands (1960) found g-values of about 2.01 and 1.94 from mitochondrial preparations and proteins of known iron content when EPR spectra from samples reduced with substrate or dithionite were taken below 150 K. In view of low flavin content and the absence of paramagnetic metal ions other than those of iron, it was proposed that the EPR signal arose from the iron.

Since electronic states of isolated iron ions would not be expected to have a g-value of 1.94, this suggestion met with some skepticism until a broadening (due to unresolved hyperfine splitting) was observed in the EPR signal from protein isolated from *Azobacter vinelandii* grown in a medium enriched in ^{57}Fe ($I = 1/2$) (Shetna et al., 1964). (It is now known that ^{57}Fe can be introduced nonbiologically.) The origin of the unusual g-value will be described after presentation of other properties of this class of proteins.

The iron-sulfur proteins which do not contain other prosthetic groups have been classified into four types: the rubredoxins, the "plant-type" ferredoxins, the "bacterial-type" ferredoxins, and the high-potential iron proteins (HiPiP) (see also Chap. 2-2 h, i). "Plant" refers to the fact that this type of ferredoxin was originally isolated from chloroplasts, but the defining property is the presence of just two iron and two inorganic sulfur atoms per protein molecule. The "bacterial" ferredoxins are those which contain more than two iron and sulfur atoms per protein molecule - usually two clusters of four each - a feature shared with HiPiP. All the ferredoxins have the unusual 1.94-1.96 g-value when reduced, and have "labile sulfur" which is unstable toward acid pH, giving off the sulfur as H_2S (a reaction not characteristic of cysteinyl sulfur and hence not characteristic of rubredoxins); and all, in the reduced form, are one-electron reducing agents with midpoint redox potentials in the range -250 to -450 mV. The high-potential iron proteins, which also contain labile sulfur but are somewhat more stable toward acid than ferreoxins, are strongly oxidizing (e.g. HiPiP from *Chromatium* has a midpoint redox potential of +350 mV), diamagnetic when reduced, and paramagnetic (g-value about 2) when oxidized (Antanaitis and Moss, 1975). Most, although not all, of the information to be given in the remainder of this section comes from the plant-type ferredoxins, with molecular weights in the range 11,000 to 24,000 daltons, which contain at least four cysteine residues per protein molecule, and which when reduced have EPR spectra characterized by $1.89 \leq g_x \leq 1.93$, $1.94 \leq g_y \leq 1.96$, $2.00 \leq g_z \leq 2.05$. Sequence information, but no X-ray structure, is available for the plant ferredoxins; cysteine residues are present; disulfide bridges are not; titration of the cysteine with heavy metals causes denaturation.

The "resting", oxidized form, gives no EPR signal and a paramagnetic susceptibility is not found at cryogenic temperatures. At room temperature, there is measurable paramagnetism, but too small to arise from any of the spin states of the uncoupled ferric ion. Furthermore, large contact shifts are observed in the proton magnetic resonance spectra (Poe et al., 1970; Poe et al., 1971). These facts, and other evidence to be presented, implicate antiferromagnetic coupling of the iron atoms in the oxidized form of the protein (Gibson et al., 1966).

It is well known from the quantum mechanics of systems containing two or more electrons that changes in the electrostatic energies usually occur when the pairing of the electron spins changes. Consequences of this are, for example, the depression in energy of spin-triplet states as compared with the corresponding singlets, and the more general multiplicity-energy relation expressed by Hund's first rule. If a pair of atoms, i and j, not necessarily adjacent, shares electrons, the dependence of the electrostatic energy upon the spin orientations can give rise to strong spin ordering forces which result in ferromagnetism and antiferromagnetism. It is convenient to express the Hamiltonian term for this spin-correlation effect

$$H_{\text{exchange}}(i,j) = -2J\underline{S}_i \cdot \underline{S}_j$$

where J is an "exchange integral". In the Hamiltonian just given, it appears that there is a *direct* coupling between $\underline{S}_i$ and $\underline{S}_j$, and there can be, but H_{exchange} does not express this. When J is positive, the system orders ferromagnetically, and when J is negative, antiferromagnetically;

ferromagnetism:	$\uparrow$ $\uparrow$ S_i S_j	$\underline{S}_i \cdot \underline{S}_j > 0$	$E_{\text{exchange}} < 0$;
$J > 0$	ground state		
antiferro-magnetism:	$\uparrow$ $\downarrow$ S_i S_j	$\underline{S}_i \cdot \underline{S}_j < 0$	$E_{\text{exchange}} < 0$.
$J < 0$	ground state		

Consider first the system consisting of two antiferromagnetically coupled high-spin ferric ions. The spin momenta add to form states of total spin $S = 0, 1, 2, 3, 4, 5$, with $S = 0$ for the ground state. From $2\underline{S}_1 \cdot \underline{S}_2 = S(S+1) - S_1(S_1+1) - S_2(S_2+1)$, with $S_1 = S_2 = 5/2$, it can be shown that the difference in energy between states differing by one unit of spin is

$$\Delta E^{\text{exchange}}_{S+1,\,S} = 2|J|(S+1).$$

The splitting of the first three levels is then

$$\begin{array}{ll} 2 \text{———} & \\ & \} \; 4|J| \\ 1 \text{———} & \\ 0 \text{———} & \} \; 2|J| \\ S & \quad \Delta E. \end{array}$$

This system exhibits a "spin-only" paramagnetism with the magnetic moment of the excited state with spin S given by $\mu = 2\beta\sqrt{S(S+1)}$. The paramagnetic susceptibility, as a function of temperature and J, is calculated in the usual way from the partition function. Thus, for the case of spinach ferredoxin, comparison with the susceptibility data shows J to be $-182\ \text{cm}^{-1}$ (Dunham et al., 1971b). Since kT at room temperature is about $200\ \text{cm}^{-1}$, the excited states are poorly populated, and what little paramagnetism there is comes largely from the first excited state.

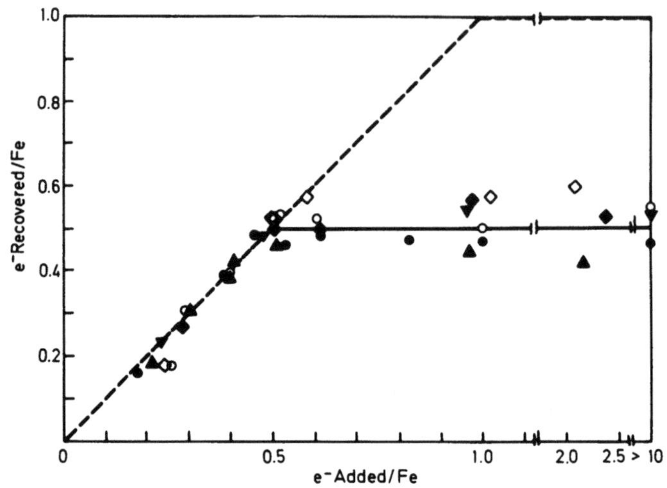

Fig. 5-1. Titration of six iron-sulfur proteins of the plant ferredoxin type with dithionite as monitored by EPR at $g = 1.94$; according to Orme-Johnson and Beinert (1969). The symbols represent putidaredoxin (▼); spinach ferredoxin (▲); bovine (◆) and pig (o) adrenal iron-sulfur proteins; and *Azobacter* iron-sulfur proteins I (O) and II ( ); methyl viologen was added to the adrenal protein. *Ascending dashed line:* the theoretical curve for 1 electron recovered in the EPR spectrum per electron added; *horizontal dashed line:* the expected curve when 2 electrons are required to reduce a given protein maximally; *horizontal full line:* the case in which 1 electron (0.5 electrons per iron atom) is needed to reduce maximally the iron proteins

Reduction titration (e.g. with dithionite) shows that a maximum of one electron can be transferred per two iron atoms of the iron-sulfur sites, as in Figure 5-1 (Orme-Johnson and Beinert, 1969) where data from six iron-sulfur proteins of the plant (ferredoxin) type show a linear relation between EPR signal and dithionite added. Taken together, the Mössbauer, EPR and susceptibility data indicate antiferromagnetic coupling of high-spin ferric, high-spin ferrous pairs in the reduced form. The spin-Hamiltonian for this site can be written, apart from hyperfine interactions,

$$H_s = V_1 + V_2 - 2J\underline{S}_1 \cdot \underline{S}_2 + D_1[S_{1z}^2 - \tfrac{1}{3}S_1(S_1+1)] + D_2[S_{2z}^2 - \tfrac{1}{3}S_2(S_2+1)]$$

$$+ g\beta S_z H_z,$$

where V_1 and V_2 are the ligand field terms, which leave both irons with nondegenerate ground states; the third term is the antiferromagnetic coupling interaction, with ion 1 taken as ferric, $S_1 = 5/2$, and ion 2 taken as ferrous, $S_2 = 2$; the following two terms are zero-field splittings (negligibly small in this case except at temperatures below the boiling point of helium); and the last term is the Zeeman interaction. Figure 5-2 (Dunham et al., 1971a) shows an energy level diagram for this system, but not to scale.

Fig. 5-2. Energy level diagram for high-spin ferrous ion antiferromagnetically spin-coupled to high-spin ferric ion. Ferrous atom is in a distorted tetrahedral ligand field. Figure drawn to scale for $J_{d_{z^2}} = -80\ cm^{-1}$, $J_{d_{x^2-y^2}} = -175\ cm^{-1}$, $\Delta = 400\ cm^{-1}$. Energy of ferric ion not to scale. Zerofield and Zeeman splittings are omitted. (From Dunham et al., 1971b)

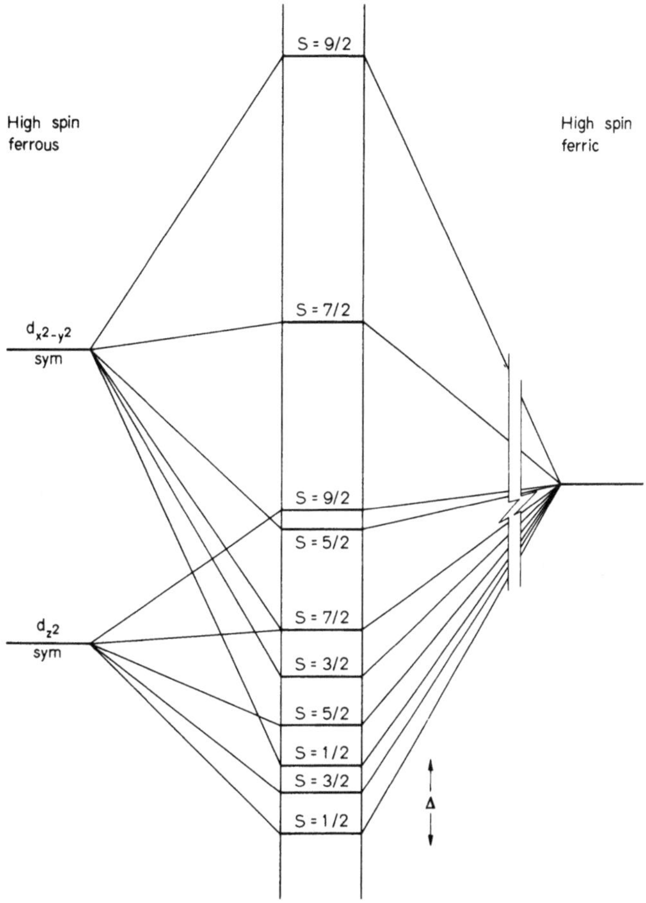

The electric field gradient tensor and hyperfine tensor obtained with ^{57}Fe Mössbauer data from the reduced protein indicate that d_{z^2} is the lowest orbital state for the ferrous ion and that the next level (taken to be $d_{x^2-y^2}$) lies about 400 cm^{-1} above d_{z^2} for S = 1/2. The term $-2J\underline{S}_1 \cdot \underline{S}_2$ can be treated as a perturbation coupling the ferric and ferrous ground states provided $|J| \ll \Delta_{ligand\ field}$, a condition which is found to be marginally met when the susceptibility measurements are analyzed. Proceeding as before, but now with $S_2 = 2$, we find for the first three d_{z^2} levels:

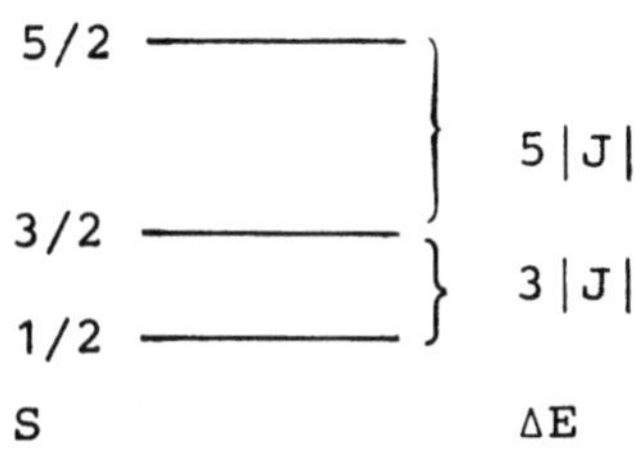

Here, however, in setting up the partition function, one must take into account the orbital contribution from the ferrous ion. The orbital contribution to the ground (d_{z^2}, S = 1/2) state is included when the observed g-values are used to calculate the Zeeman energy. Similarly the g-values for the states d_{z^2}, $S \geq 3/2$ are required; these are given by

$$g|S| = g_1 \frac{\underline{S}_1 \cdot \underline{S}}{|S|} + g_2 \frac{\underline{S}_2 \cdot \underline{S}}{|S|} ,$$

where g_1 (from the high-spin ferric S-state ion) is assumed isotropic and given the value 2.019 (Gibson et al., 1966). g_2 is obtained from the observed g-values by applying the formula just given to the S = 1/2 ground state (e.g. $\underline{S}_1 \cdot \underline{S} = \underline{S}_1 \cdot (\underline{S}_1 + \underline{S}_2) = |S_1|^2 + \frac{1}{2}\{|S|^2 - |S_1|^2 - |S_2|^2\} = 7/4$):

$$g_i = 1/3(7g_{1i} - 4g_{2i}), \quad i = x, y, z.$$

For spinach ferredoxin $g_x = 1.89$, $g_y = 1.96$, $g_z = 2.05$ and $g_{2x} = 2.12$, $g_{2y} = 2.07$, and $g_{2z} = 2.00$. The susceptibility data, considered in conjunction with the partition function, give J for d_{z^2} to be in the range -80 to -100 cm^{-1}. The first excited state (S = 3/2) of the orbitally lowest antiferromagnetically coupled spin manifold is seen to be well-populated at room temperature.

The question then arises about the antiferromagnetically coupled manifold arising from the first excited electronic state of the ferrous ion, mentioned earlier to be $d_{x^2-y^2}$ at about 400 cm^{-1} above d_{z^2} for S = 1/2. The IR spectra of reduced parsley and spinach ferredoxin and rubredoxin, show weak absorption bands at 4000 and 6900 cm^{-1} (Eaton et al., 1971). The high optical activity of these bands indicates that they are transitions from d_{z^2} to d_{yz} and d_{xz} rather than to $d_{x^2-y^2}$. This data supports the interpretation of the Mössbauer spectra given earlier, which places the lowest three levels as follows:

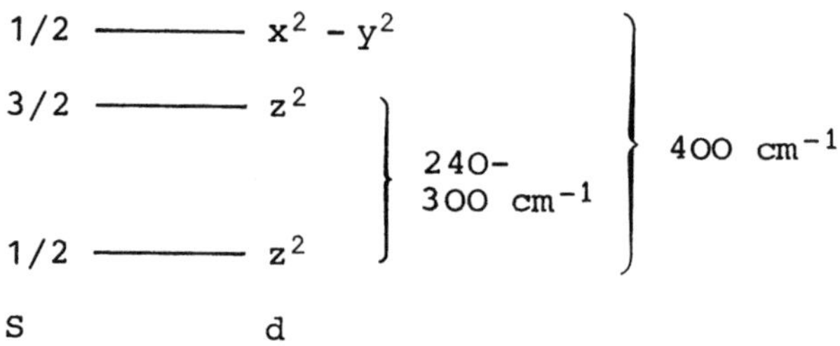

One has to take susceptibility and EPR data at rather high temperatures to determine whether or not $d_{x^2-y^2}$ does contribute. However, there is unlikely to be an EPR signal at the required temperatures on the basis of what has been observed so far for iron-sulfur proteins (Beinert, 1972); the spectrum from spinach ferredoxin is poor even at 104 K.

Proton magnetic resonance contact shifts from the reduced protein show a temperature dependence which is anomolous for an isolated magnetic center but can be explained in terms of the coupled model. Many of the considerations just presented for the quantification of the magnetic susceptibility apply to estimation of the contact shifts.

The origin of the g-values for reduced iron-sulfur proteins is now readily seen. They are given by

$$g_i = g_1 \frac{\underline{S}_1 \cdot \underline{S}}{S(S+1)} + g_2 \frac{\underline{S}_2 \cdot \underline{S}}{S(S+1)},$$

where $g_1 \approx 2.02$ and isotropic, and the g_2 values are obtained in the usual way for a hole of spin 2, spin-orbit coupling constant $\lambda \approx 80\ \text{cm}^{-1}$, from the energy level diagram:

d_{xy} ————

d_{xz} ———— $6900\ \text{cm}^{-1}$

d_{yz} ———— $4000\ \text{cm}^{-1}$

$d_{x^2-y^2}$ ————

d_{z^2} ———— $0\ \text{cm}^{-1}$.

That is $g_{2_z} = 2$, $g_{2_i} = 2(1 + 3\lambda/\Delta_{iz})$.

Substitutions of ^{57}Fe, ^{33}S and ^{77}Se for the more abundant iron and sulfur isotopes that do not have nuclear paramagnetism have enabled EPR spectroscopy, through hyperfine interactions, to demonstrate that both iron atoms and labile sulfur are involved in the magnetic sites. These isotope exchanges are carried out by acidification and replacement, and result in analogs that are enzymatically active. The biological incorporation of ^{33}S (bacteria grown in an isotopically enriched medium) has further enabled EPR to show that at least one cysteine or methionine sulfur is also at the active site. While ENDOR measurements on ^{57}Fe substituted proteins have demonstrated that the two irons are electronically inequivalent in the reduced form, in the oxidized form the two ferric ions are in almost identical electronic environments. The reader can find further details of these experiments, related measurements, analyses, and model considerations in: Dunham et al., 1971a; Dunham et al., 1971b; Fritz et al., 1971; Palmer et al., 1971.

An eight atom structure has been proposed which is consistent with the results of these investigations. In this structure, there are two labile sulfur atoms which bridge the two iron atoms.

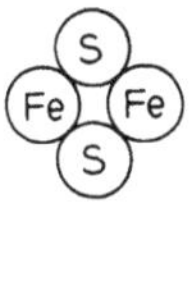

A pair of cysteinyl-sulfur ligands are also bound to each iron in the proposed site, tetrahedrally disposed with respect to the labile sulfur pair.

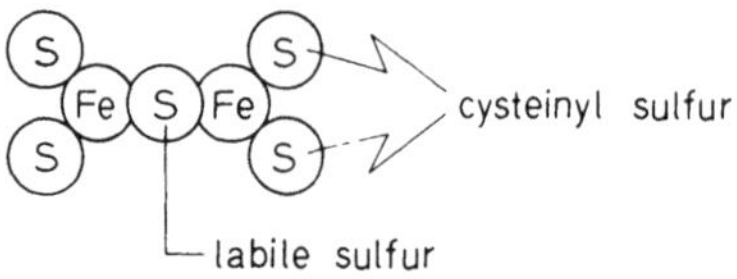

X-ray diffraction-determined structures are not yet available for any iron-sulfur proteins of the plant type, which is the type largely responsible for the data leading to the complex just pictured. The X-ray structures (see Chap. 2-2, h,i) of rubredoxin (Watenpaugh et al., 1973), which does not contain acid-labile sulfur, and multi-iron-sulfur proteins of bacterial origin (Carter et al., 1971; Sieker et al., 1972) do not support in detail this structure; nor do model compounds which have certain electronic properties in common with those of the iron-sulfur proteins (Carrell et al., 1977; Herskovitz et al., 1972). However, in these cases, the X-ray-determined geometry about each iron does have low symmetry. The bacterial proteins and the model complexes have approximately cubical Fe_4S_4 cores ("cubane clusters") in which iron and sulfur atoms are at alternate corners.

The techniques which have led to the electronic description outlined earlier of the plant-type iron-sulfur system are also being applied to the more complex bacterial-type proteins (Mathews et al., 1974). Further information is becoming available from resonance Raman spectroscopy (Tang et al., 1975).

3. Molybdenum

Molybdenum plays an essential role in the biological oxidations catalyzed by such flavoprotein enzymes as xanthine oxidase, nitrate reductase, and aldehyde oxidase; it also appears to be functional in nitrogenases, which contain more iron than molybdenum. The known valence states of molybdenum together with their electronic configurations and spin when bound in complexes are listed in Table 5-1. Mo(VI) is necessarily diamagnetic; Mo(IV) and Mo(II) are usually found to be diamagnetic. The paramagnetic valences, V and III, have little orbital angular momentum associated with them. The low magnetic moments are not surprising since the coordination symmetry of molybdenum compounds, established from X-ray structure determinations, is frequently low. It is interesting to note that both quadrivalent and quinquevalent molybdenum have the high coordination number of eight in the octacyanides $[Mo(CN)_8]^{3-}$ and $^{4-}$, but the symmetry of these compounds is not correspondingly high.

Table 5-1. Valence states, electronic configurations and spin states of molybdenum

Ions	Ground configuration	Spin in complexes
Mo(VI)	$4d^0$	0
Mo(V)	$4d^1$	1/2
Mo(IV)	$4d^2$	0, 1(?)
Mo(III)	$4d^3$	3/2
Mo(II)	$4d^4$	0

Mo(II) is highly reductive and unlikely to have a physiological role. This suggests that Mo(III) is also not biochemically functional because the enzymatic state which exhibits an electron paramagnetic resonance spectrum no longer does so when the enzyme is reduced by substrate or other electron donors (Bray et al., 1961). Thus, Mo(V) is inferred to be the EPR active valence state which participates with, probably, both Mo(IV) and Mo(VI) in enzymatic processes.

Electron paramagnetic resonance studies of small Mo(V) complexes in crystals and solutions show g tensors which are axial or slightly rhombic with $g_{||}$ in the range 1.95 to 2.01 and $g_{\perp}$, 1.92 to 2.00. The even isotopes (94, 96, 98), which have no nuclear spin and hence no hyperfine interactions, constitute about 75% of naturally occurring molybdenum. The odd isotopes, 95 and 97, occur with a natural abundance of 15.7% and 9.5% respectively, have a nuclear spin of 5/2, and give rise to six hyperfine lines. Because of the mixed hyperfine behavior, EPR spectra from complexes made with natural molybdenum can be difficult to interpret. To avoid this, much work has been done with isotopically enriched molybdenum (up to 97% ^{95}Mo). For ^{95}Mo, the A-tensor has principal values in the ranges 0.004 to 0.008 cm^{-1} for A_z and 0.002 to 0.004 cm^{-1} for A_x and A_y. The following observations are based upon the relatively small number of Mo(V) complexes from which EPR data is available. In complexes coordinating through sulfur and in the octacyanide complex, the g-values are closer to that of the free electron, and the A-values are lower than in other complexes. The octacyanide complex is distinguished by its unusual symmetry and the covalent character of the coordinate links. The symmetry of the sulfur-containing complexes is probably not special, but sulfur does behave like an unsaturated ligand. Metal electron delocalization is then seen to account for the higher g-values and lower A-values in Mo(V) thiol complexes (Meriwether et al., 1966).

EPR spectra from the molybdenum-containing enzymes exhibit a region in which the parameters are similar to those from Mo(V)-thiol complexes and $[Mo(CN)_8]^{3-}$. Proof that this absorption is associated with molybdenum was obtained in a classic demonstration of the use of isotopic labeling to interpret EPR spectra (Bray et al., 1967). A cow was injected with ^{95}Mo and xanthine oxidase was prepared from the milk collected over the next several days. EPR spectra from the normal enzyme and from the ^{95}Mo-

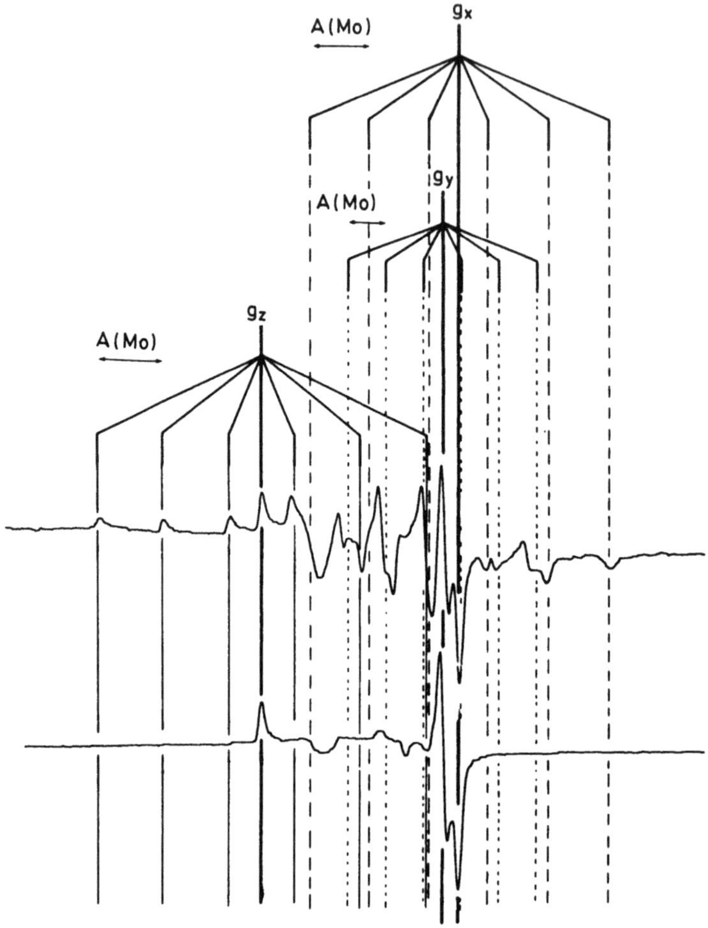

Fig. 5-3. EPR signal from xanthine oxidase: *lower trace:* normal enzyme; *upper trace:* ^{95}Mo enzyme. The isotopes with I = 0 give the three main lines belonging to g_x, g_y and g_z. In the ^{95}Mo protein, each of these lines is replaced by six equally spaced hyperfine lines. (From Bray and Meriwether, 1966)

enzyme are shown in Figure 5-3. In the latter record, the hyperfine interactions characteristic of the odd molybdenum isotope can be clearly resolved. Xanthine oxidase is inactivated by the sulfhydryl reagent iodoacetamide with substantial loss of flavin and significant alteration in the EPR spectrum. The g-values are the lowest of all enzyme-associated Mo(V) signals, though on the high side for nonsulfur complexes. (A-values from ^{95}Mo are apparently small but thus far unresolved.) Taken together, the several kinds of evidence favor one or more sulfur-molybdenum links in the native enzyme, and indicate that other ligands either tend to withdraw metal electrons or else are part of a metal-binding site of low symmetry.

The reader is referred to Bray and Swann (1972) for a review of molybdenum-containing enzymes which treats the electronic aspects of metal binding, to Bray (1975) for detailed discussion of the molecular and catalytic properties and a survey of genetic studies of molybdenum-iron sulfur flavin hydroxylases, and to Wentworth (1976) for a review of the aqueous chemistry of the higher oxidation states of molybdenum as it relates to current theories of reaction mechanisms in molybdenum enzymes.

Chapter 6

Electronic Structures and Properties[5]

1. Atomic Orbitals

To a first approximation, the valence electrons of free atoms and ions are taken to move in a spherically symmetric potential, and interactions between them are neglected. The resulting one-electron wave functions can be written as the product

$$\psi_{n\ell m} = R_n^{\ell}(r)Y_{\ell}^{m}(\theta,\phi)$$

where n is the "principal" quantum number; ℓ, the orbital angular momentum quantum number; and m, the quantum number characterizing the "z-component" of orbital angular momentum. The energy depends largely upon n, less upon ℓ (not at all for a Coulomb potential), and not upon m. Y_{ℓ}^{m} are the spherical harmonics, functions which in turn are separable into a θ-dependent part and a ϕ-dependent part, $e^{im\phi}$. The orbital angular momentum operator is denoted $\underline{L}$ and the orbital angular momentum is specified (as well as is possible) by the eigenvalues of $\underline{L}^2$ and L_z (or any other single component), i.e.,

$$\underline{L}^2Y_{\ell}^{m} = \ell(\ell + 1)\hbar^2Y_{\ell}^{m}, \quad \ell = 0 \text{ or any positive integer;}$$

$$L_zY_{\ell}^{m} = m\hbar Y_{\ell}^{m}, \quad -\ell \leq m \leq \ell, \text{ m integral.}$$

In the remainder of the text, angular momentum will be measured in units of $\hbar$ in order to avoid continually writing $\hbar$ on the right-hand sides of relations such as the preceding two. Closed form approximations to the radial part have been used (e.g. orbitals of the form given by Zener (1930) and Slater (1930)) and numerical functions are available for many atoms and ions (Clementi, 1964; Clementi et al., 1962; Hartree, 1957; Richardson et al., 1962; Richardson et al., 1963; Watson, 1960a,b).

The atomic orbitals of concern in biological structures are those with ℓ values of 0, 1, and 2, respectively designated by the symbols s, p, and d. It is useful to describe these as real functions in space, and in only a few cases in the subject area of concern here is it incorrect to do so. The angular dependence of the real p and d orbitals is schematized in Figure 6-1, wherein are also given the functional and symbolic forms customarily em-

---

[5] The reader is assumed to have some familiarity with quantum mechanics

Table 6-1. Representations of $\underline{L}$ with respect to the real p and d bases

	p_x	p_y	p_z		
p_x	0	$-i\underline{k}$	$i\underline{j}$		
p_y	$i\underline{k}$	0	$-i\underline{i}$		
p_z	$-i\underline{j}$	$i\underline{i}$	0		
	$d_{x^2-y^2}$	d_{z^2}	d_{xy}	d_{yz}	d_{xz}
$d_{x^2-y^2}$	0	0	$-i2\underline{k}$	$i\underline{i}$	$i\underline{j}$
d_{z^2}	0	0	0	$i\sqrt{3}\,\underline{i}$	$-i\sqrt{3}\,\underline{j}$
d_{xy}	$i2\underline{k}$	0	0	$i\underline{j}$	$-i\underline{i}$
d_{yz}	$-i\underline{i}$	$-i\sqrt{3}\,\underline{i}$	$-i\underline{j}$	0	$i\underline{k}$
d_{xz}	$-i\underline{j}$	$i\sqrt{3}\,\underline{j}$	$i\underline{i}$	$-i\underline{k}$	0

ployed in various calculations. The ket notation $|m\rangle$, e.g. $d_{z^2} = Y_2^0 = |0\rangle$, is a widely used formalism to depict states, in these cases eigenstates of L_z. (The value of ℓ can be explicitly written, e.g. $d_{z^2} = |2,0\rangle$, but it is not necessary to do this when only the manifold of m states corresponding to a given ℓ is being considered.) Note that most of the real orbitals are not eigenfunctions of L_z, e.g.

$$L_z d_{x^2-y^2} = L_z \frac{1}{\sqrt{2}}(|2\rangle + |-2\rangle) = \frac{1}{\sqrt{2}}(L_z|2\rangle + L_z|-2\rangle)$$

$$= \frac{2}{\sqrt{2}}(|2\rangle - |-2\rangle) = 2id_{xy}.$$

Those real orbitals which are eigenfunctions of L_z have $m = 0$, e.g.

$$L_z p_z = L_z|0\rangle = 0|0\rangle = 0.$$

The matrix elements of orbital angular momentum among the real p and d orbitals are listed in Table 6-1. The x and y momentum operators are readily handled as linear combinations of the "raising" operator (non-Hermitian)

Fig. 6-1. Angular dependence of real p and d orbitals. The symbols e_g and t_{2g} refer to irreducible representations of the octahedral group O_h to which the functions belong. The symbols $d\gamma$, $d\varepsilon$ and Γ_3, Γ_5 have the same significance as e_g, t_{2g} in other, less systematic notations

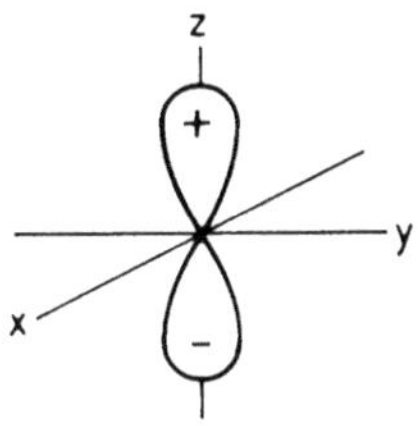

$p_z = Y_1^0$

$p_z = |0\rangle$

$p_z = \sqrt{\frac{3}{4\pi}} \cos\Theta$

$p_z = \sqrt{\frac{3}{4\pi}} \frac{z}{r}$

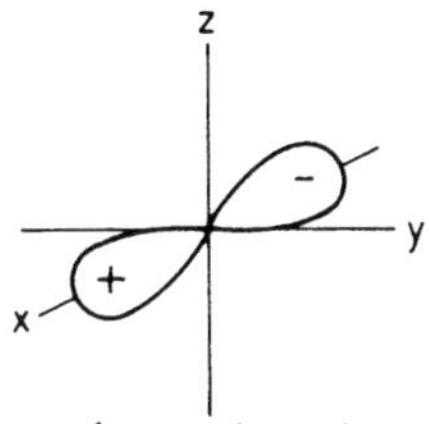

$p_x = \frac{1}{\sqrt{2}} (Y_1^1 + Y_1^{-1})$

$p_x = \frac{1}{\sqrt{2}} (|1\rangle + |-1\rangle)$

$p_x = \sqrt{\frac{3}{4\pi}} \sin\Theta \cos\phi$

$p_x = \sqrt{\frac{3}{4\pi}} \frac{x}{r}$

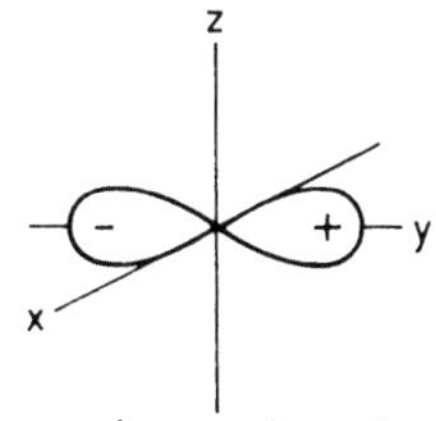

$p_y = \frac{1}{i\sqrt{2}} (Y_1^1 - Y_1^{-1})$

$p_y = \frac{1}{i\sqrt{2}} (|1\rangle - |-1\rangle)$

$p_y = \sqrt{\frac{3}{4\pi}} \sin\Theta \sin\phi$

$p_y = \sqrt{\frac{3}{4\pi}} \frac{y}{r}$

p orbitals

$e_g / d\gamma / \Gamma_3$

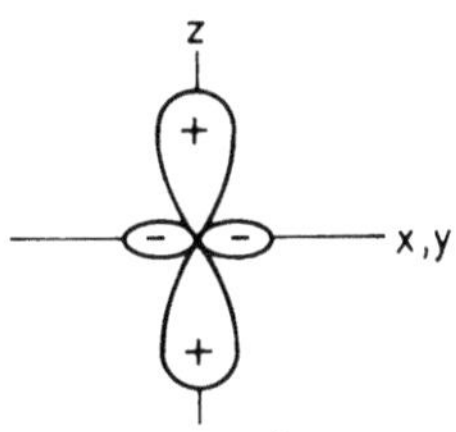

$d_{z^2} = d_{3z^2-r^2} = Y_2^0$

$d_{z^2} = |0\rangle$

$d_{z^2} = \frac{1}{4} \sqrt{\frac{5}{\pi}} (3 \cos^2\Theta - 1)$

$d_{z^2} = \frac{1}{4} \sqrt{\frac{5}{\pi}} \frac{2z^2 - x^2 - y^2}{r^2}$

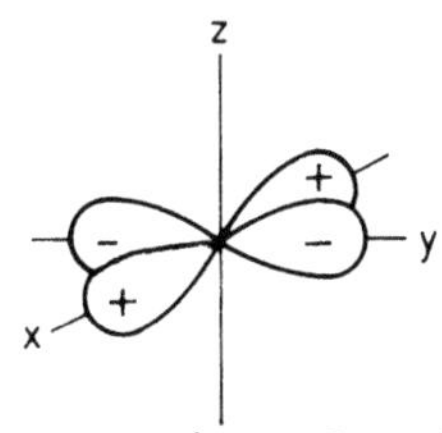

$d_{x^2-y^2} = \frac{1}{\sqrt{2}} (Y_2^2 + Y_2^{-2})$

$d_{x^2-y^2} = \frac{1}{\sqrt{2}} (|2\rangle + |-2\rangle)$

$d_{x^2-y^2} = \frac{1}{4} \sqrt{\frac{15}{\pi}} \sin^2\Theta \cos 2\phi$

$d_{x^2-y^2} = \frac{1}{4} \sqrt{\frac{15}{\pi}} \frac{x^2 - y^2}{r^2}$

$t_{2g} / d\varepsilon / \Gamma_5$

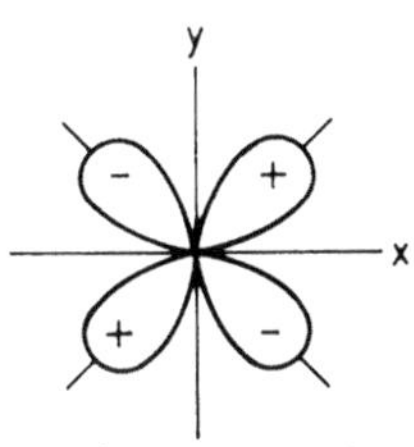

$d_{xy} = \frac{1}{i\sqrt{2}} (Y_2^2 - Y_2^{-2})$

$d_{xy} = \frac{1}{i\sqrt{2}} (|2\rangle - |-2\rangle)$

$d_{xy} = \frac{1}{2} \sqrt{\frac{15}{\pi}} \sin^2\Theta \sin 2\phi$

$d_{xy} = \frac{1}{2} \sqrt{\frac{15}{\pi}} \frac{xy}{r^2}$

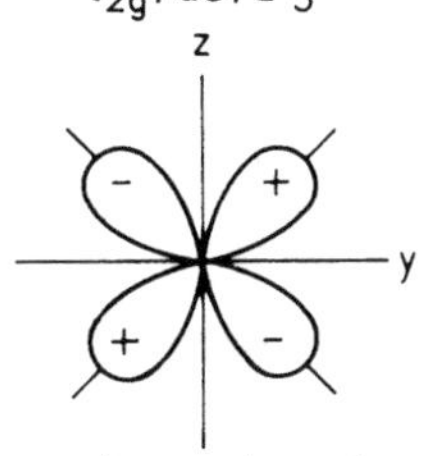

$d_{yz} = \frac{1}{i\sqrt{2}} (Y_2^1 - Y_2^{-1})$

$d_{yz} = \frac{1}{i\sqrt{2}} (|1\rangle - |-1\rangle)$

$d_{yz} = \frac{1}{2} \sqrt{\frac{15}{\pi}} \sin\Theta \cos\Theta \sin\phi$

$d_{yz} = \frac{1}{2} \sqrt{\frac{15}{\pi}} \frac{yz}{r^2}$

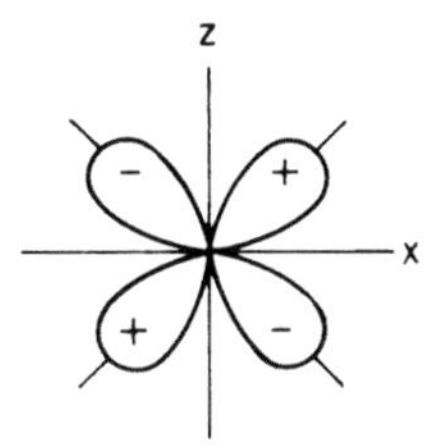

$d_{xz} = \frac{1}{\sqrt{2}} (Y_2^1 + Y_2^{-1})$

$d_{xz} = \frac{1}{\sqrt{2}} (|1\rangle + |-1\rangle)$

$d_{xz} = \frac{1}{2} \sqrt{\frac{15}{\pi}} \sin\Theta \cos\Theta \cos\phi$

$d_{xz} = \frac{1}{2} \sqrt{\frac{15}{\pi}} \frac{xz}{r^2}$

d orbitals

Fig. 6-1

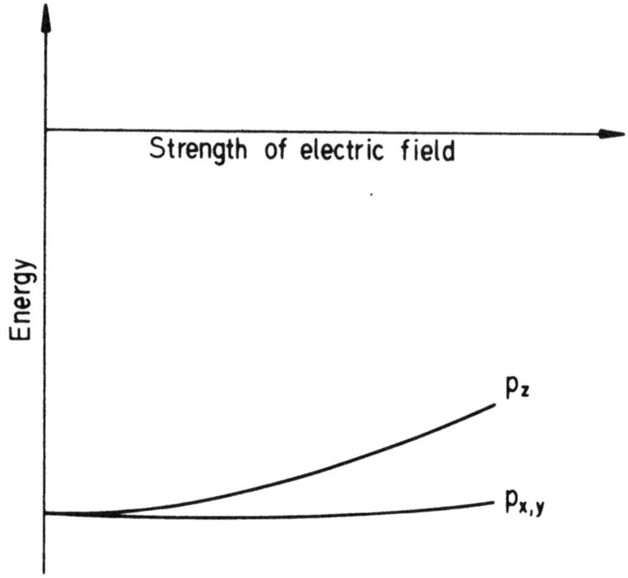

Fig. 6-2. Effect of homogeneous electric field in z-direction upon energies of atomic p states

$$L_+ \equiv L_x + iL_y$$

and the "lowering" operator (also non-Hermitian)

$$L_- \equiv L_x - iL_y$$

where

$$L_+ Y_\ell^m = \pm \sqrt{\ell(\ell+1) - m(m+1)}\, Y_\ell^{m+1} \begin{cases} + \text{ for } m < 0 \\ - \text{ for } m \geq 0 \end{cases}$$

and

$$L_- Y_\ell^m = \pm \sqrt{\ell(\ell+1) - m(m-1)}\, Y_\ell^{m-1} \begin{cases} + \text{ for } m \leq 0 \\ - \text{ for } m > 0 \end{cases}$$

(the spherical harmonics Y_ℓ^m are taken positive), namely

$$L_x = \frac{1}{2}(L_+ + L_-) \text{ and } L_y = \frac{1}{2i}(L_+ - L_-) .$$

Note that the non-zero matrix elements are all wholly imaginary.

On a single atom or ion, for a given value of ℓ the "one-electron" atomic orbitals have the same energy. Thus, the manifold of p-orbitals is triply degenerate and the d-orbitals, five-fold degenerate. When the atom or ion is subject to an electrostatic field, differences in energy among the members of an ℓ-manifold appear, giving rise to spectroscopic effects bearing the name of Stark. For example, a homogeneous electric field in the z-direction affects atomic p-levels, as shown in Figure 6-2. The fields

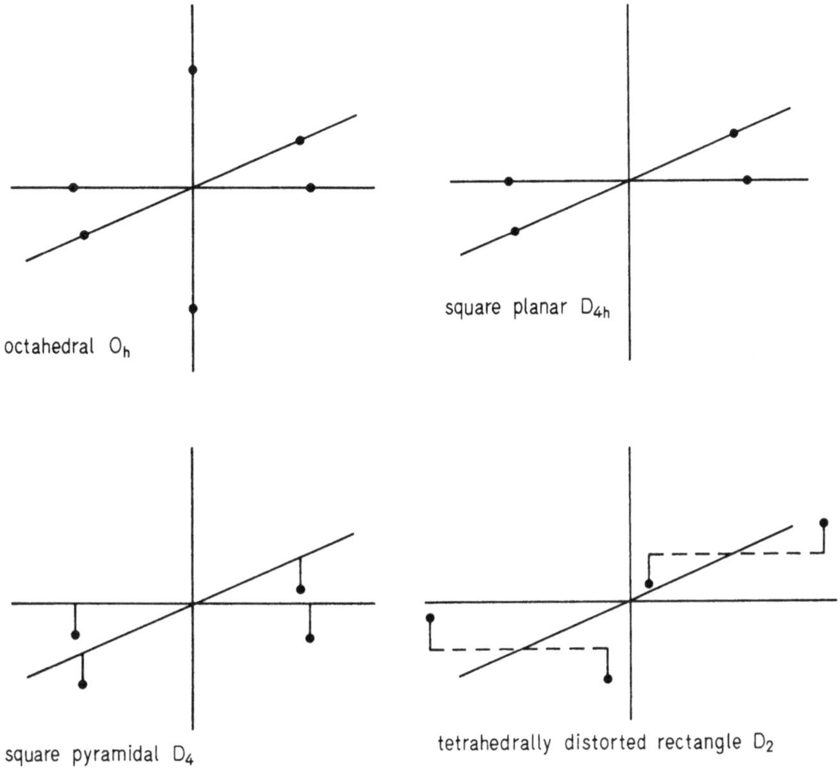

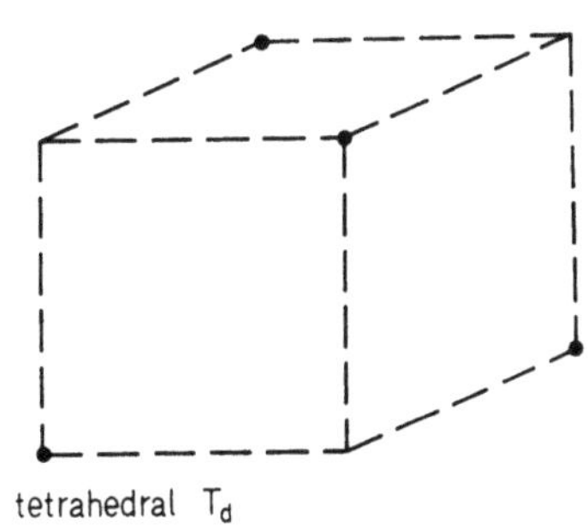

Fig. 6-3. Figures depicting symmetries, with names and group symbols. In the figure for O_h, the points are at the centers of the faces of a cube; in the figure for T_d, the points are at diagonally opposite corners of a cube; O_h and T_d are said to have cubic symmetry

at lattice points in a crystal or at the center of the coordination "sphere" of a metal complex are specified by symmetry and, of course, strength. The symmetry is defined by the set of operations (e.g. rotations and reflections) which leave the field, or equivalently the electrostatic potential, unchanged. Figures depicting symmetries of interest here, the names of the symmetries and their group symbols are in Figure 6-3. The nomenclature of group representations is used in this book, but knowledge of group theory is not required to follow the text. In Figure 6-4, the effects of changing symmetry upon the d-orbitals are followed. Here we suppose negative charge in the d-orbitals and

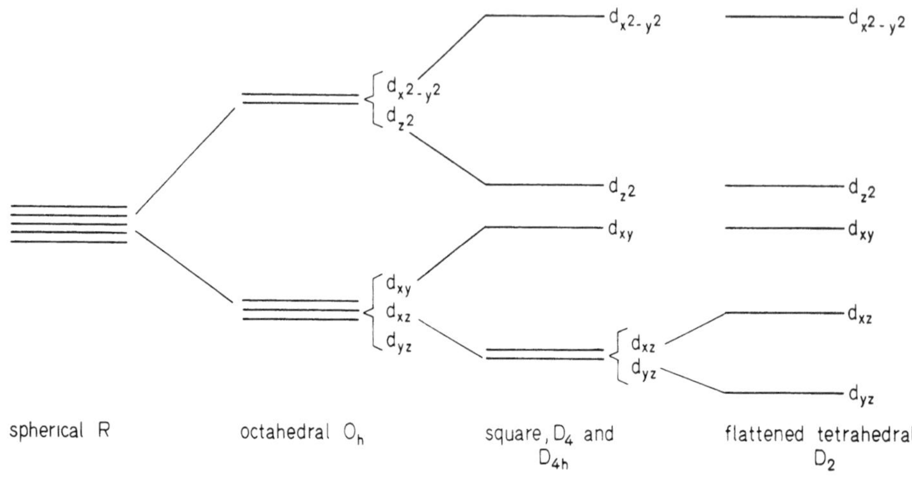

Fig. 6-4. Effects of changing symmetry upon d-levels

at the vertices of the figures such that the repulsive energy of an orbital lobe pointing at a vertex is greater (more positive) than for a lobe concentrated off-axis. Bleaney and Stevens (1953) outline methods, not based upon the use of group theory, for finding the splittings which are caused by the "crystal field"; further exposition appears in Dunn et al. (1965). The application of group representation theory to this problem is discussed by Tinkham (1964) and treated comprehensively by Griffith (1961).

2. Spin States

Each "one-electron" spatial orbital described above accommodates two electrons which must differ in spin state (to satisfy the Pauli exclusion principle) but, in first approximation, do not interact otherwise. The spin angular momentum operator is denoted $\underline{S}$ and this momentum is specified as well as is possible by the eigenvalues of $\underline{S}^2$ and S_z (or any other single component), i.e.

$$\underline{S}^2 |S,M_S\rangle = S(S+1)|S,M_S\rangle$$

$$S_z|S,M_S\rangle = M_S|S,M_S\rangle$$

where, for a single electron, $S = 1/2$ and $M_S = \pm 1/2$. It is customary to denote the states $|1/2, 1/2\rangle$ and $|1/2, -1/2\rangle$ by the symbols α and β. Thus, for example, the ground state of helium could be designated $(1s(1)\alpha(1), 1s(2)\beta(2))$, where the bracketed numbers refer to electrons 1 and 2, were it not for the indistinguishability of electrons which requires a linear combination (in this case, antisymmetric) of the product spin state; that is

$$1s(1)1s(2)[\alpha(1)\beta(2) - \beta(1)\alpha(2)]/\sqrt{2}.$$

When two different spatial orbitals are occupied by two electrons, the individual spin states can combine in more ways, as follows:

Spin singlet	*Spin triplet*
one state	manifold of three states
$S = 0$	$S = 1$
$M_S = 0$, $[\alpha(1)\beta(2) - \beta(1)\alpha(2)]/\sqrt{2}$	$M_S = 1$, $\alpha(1)\alpha(2)$
	$M_S = 0$, $[\alpha(1)\beta(2) + \beta(1)\alpha(2)]/\sqrt{2}$
	$M_S = -1$, $\beta(1)\beta(2)$
antisymmetric with respect to interchange of electrons	symmetric with respect to interchange of electrons

The values of M_S can be seen immediately upon use of S_z as the operator sum $S_{1z} + S_{2z}$. In order to demonstrate the values of S for these spin state functions, one first expands $\underline{S}^2$,

$$\underline{S}^2 = (\underline{S}_1 + \underline{S}_2)\cdot(\underline{S}_1 + \underline{S}_2) = \underline{S}_1^2 + \underline{S}_2^2 + 2(S_{1x}S_{2x} + S_{1y}S_{2y} + S_{1z}S_{2z}).$$

When, as here, the individual states are eigenstates of S_{iz}, it is convenient to carry out operations involving x and y components by means of the substitutions

$$S_x = \tfrac{1}{2}(S_+ + S_-) \text{ and } S_y = \tfrac{1}{2i}(S_+ - S_-)$$

where S_+ and S_- are "raising" and "lowering" operators (non-Hermitian) with the properties

$$S_+\alpha = 0, \quad S_+\beta = \alpha, \quad S_-\alpha = \beta, \quad \text{and } S_-\beta = 0.$$

One then arrives at

$$\underline{S}^2 = \tfrac{3}{2} + S_{1+}S_{2-} + S_{1-}S_{2+} + 2S_{1z}S_{2z}$$

with the eigenvalues 0 for the singlet state and 2 for the triplet states.

For a system of n electrons, the maximum value of S is n/2, the minimum value is 0 if n is even and 1/2 if n is odd, and the manifold of M_S values for a given S goes from -S to S by integral steps. The number, $2S + 1$, of states in such a manifold is called the (spin) "multiplicity", e.g.

$S = 0$, multiplicity of 1 (a singlet), $M_S = 0$

$S = \frac{1}{2}$, multiplicity of 2 (a doublet), $M_S = -\frac{1}{2}, \frac{1}{2}$

$S = \frac{5}{2}$, multiplicity of 6 (a sextet), $M_S = -\frac{5}{2}, -\frac{3}{2}, -\frac{1}{2}, \frac{1}{2}, \frac{3}{2}, \frac{5}{2}$.

The raising and lowering spin operations are given by

$$S_+|S,M_s\rangle = \sqrt{S(S+1) - M_s(M_s+1)}\ |S,M_s+1\rangle$$

and

$$S_-|S,M_s\rangle = \sqrt{S(S+1) - M_s(M_s-1)}\ |S,M_s-1\rangle\ .$$

In free atoms and ions, the energies depend upon S through spin-orbit coupling (see Chap. 6-3) but not upon M_S. For atoms and ions in crystals and molecules, when the multiplicity is greater than 2, the energy can depend upon $|M_S|$ even in the absence of an applied magnetic field. A Hamiltonian term expressing this "zero-field splitting" can be written

$$D[S_z^2 - \tfrac{1}{3}S(S+1)]$$

where D is a parameter which depends upon the (spatial) orbitals. Thus, for a quintet, the states with $M_S = \pm\frac{1}{2}$ are separated from those with $M_S = \pm\frac{3}{2}$ by 2D, and from those with $M_S = \pm\frac{5}{2}$ by 6D. When significant zero-field splitting is present, the direction z is no longer arbitrary, but is determined by the environment. Anisotropy can be present about the z-direction, in which case there is another term in the Hamiltonian which contains only spin operators,

$$E(S_x^2 - S_y^2)\ .$$

These axial and rhombic quadratic spin terms account in part for the dependence of certain measurements upon the orientation of the system with respect to a polarizing magnetic field. Further discussion of D and E, and the magnetic effects of zero-field splitting, are found in Chapters 4-6 and 6-6.

Atomic nuclei have characteristic spin quantum numbers, denoted I, with z-components m_I. When $I > 0$, the nucleus has a magnetic moment

$$\underline{\mu}_n = g_n \beta_n \underline{I}$$

where β_n is the nuclear magneton and g_n, the "nuclear g-value", is a property of the isotope under consideration. The electrons and the nuclei of the same atom interact magnetically, and there can also be significant magnetic interactions of the electrons of an atom with the nuclear moments of nearby atoms, particularly when there is delocalization of electrons onto the neighbors (covalency). These "hyperfine" interactions are of three kinds: Fermi (or contact or isotropic), orbital field-nuclear dipole, and electron spin dipole-nuclear dipole (Abragam and Pryce, 1951). The Fermi interaction is present when unpaired electron spin is in an orbital with $s(\ell = 0)$ character. (Only s-orbitals are non-zero at the nucleus.) When there is unpaired spin in an open shell, closed s-shells can contribute to the Fermi interaction through "core polarization", an unbalancing of the spin distribution which results in a small unpaired spin density at the nucleus. An electron with s-character is in "contact" with the nucleus, and the resulting magnetic coupling is expressed

$$\mathcal{H}_{\text{Fermi}} = A_F \underline{I} \cdot \underline{S}.$$

The interaction of the nuclear magnetic dipole with the magnetic field arising from the orbital motion of the electrons is given by

$$\mathcal{H}_{\text{orbital field-nuclear dipole}} = P\underline{L} \cdot \underline{I}$$

where $P = 2.0023\beta g_n \beta_n \langle r^{-3} \rangle$, β is the Bohr magneton, $\langle r^{-3} \rangle$ is the expectation (average) value of r^{-3} for the orbitals under consideration, and $\underline{L}$ is summed over the electrons (Chap. 6-3). The electron spin dipole-nuclear dipole hyperfine interaction is derivable from the classical form

$$\mathcal{H}_{\text{electron dipole-nuclear dipole}} = -\frac{\underline{\mu}_e \cdot \underline{\mu}_n}{r^3} + 3\frac{(\underline{\mu}_e \cdot \underline{r})(\underline{\mu}_n \cdot \underline{r})}{r^5}$$

with the use of operator equivalents (Bleaney and Stevens, 1953) to give

$$H_{\text{electron dipole-nuclear dipole}} = P\,\xi\{L(L+1)(\underline{S} \cdot \underline{I}) - \tfrac{3}{2}(\underline{L} \cdot \underline{S})(\underline{L} \cdot \underline{I}) - \tfrac{3}{2}(\underline{L} \cdot \underline{I})(\underline{L} \cdot \underline{S})\}$$

where

$$\xi = [(2\ell + 1) - 4S]/S(2\ell - 1)(2\ell + 3)(2L - 1)$$

(L is discussed in the next section.)

When $I > \frac{1}{2}$, the nucleus has a quadrupole moment. At the nucleus there can be an electric field gradient arising from asymmetric electronic distribution in the valence shell, in which case the quadrupole moment is coupled to the gradient according to

$$\mathcal{H}_{\text{quadrupole}} = Q[I_z^2 - \tfrac{1}{3}I(I+1) + \eta(I_x^2 - I_y^2)]$$

where Q is a parameter which incorporates both the nuclear and electronic properties and η allows for the electric field gradient to be of lower than axial symmetry.

3. Transition Metal Ions

The ground configurations of the first series of transition metal ions were given in Table 1-1. In the absence of a binding structure, the energy of the isolated (free) ion is independent of orientation and hence the total angular momentum

$$\underline{J} = \underline{L} + \underline{S}$$

commutes with the Hamiltonian and is a constant of motion. $\underline{J}$ is specified as well as is possible by the eigenvalues of $\underline{J}^2$ and J_z (or any other single component), i.e.

$$\underline{J}^2 |j,m_j\rangle = j(j+1)|j,m_j\rangle \text{ , j zero or any positive half-integer or integer}$$

$$J_z|j,m_j\rangle = m_j|j,m_j\rangle \text{ , } m_j = -j, -j+1, \ldots, j-1, j.$$

For the ions of concern here, Russell-Saunders coupling applies - that is, the coupling of electron spin with orbital angular momentum is weaker than the coupling of the spins with each other and/or the coupling of the orbital moments with each other. In this case, the total orbital angular momentum

$$\underline{L} = \sum_{\text{electrons}} \underline{L}_i$$

and the total spin angular momentum

$$\underline{S} = \sum_{\text{electrons}} \underline{S}_i$$

are fixed in magnitude but precess about $\underline{J}$

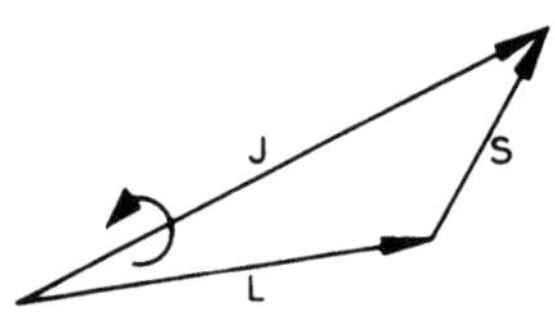

As before, the magnitudes of $\underline{L}$ and $\underline{S}$ are $\sqrt{L(L+1)}$ and $\sqrt{S(S+1)}$ respectively. The reader is encouraged to consult a standard text on atomic physics for an exposition of the addition of $\underline{L}$ and $\underline{S}$ vectors to form various $\underline{J}$. For present purposes, it suffices to note that the free ion ground states are characterized primarily by the values of (the total) L and (the total) S, and that the number of values of j is equal to the spin multiplicity 2S + 1. The manifolds of states are denoted by the spectroscopic term symbols

$$^{2S+1}L_j$$

where the value of L is signified by a capital letter with the correspondence S ⟷ 0, P ⟷ 1, D ⟷ 2, etc. For example, a single d-electron has $S = \frac{1}{2}$, and L = 2 and the term symbols are $^2D_{3/2}$ or $^2D_{5/2}$ for $j = \frac{3}{2}$ or $\frac{5}{2}$ respectively. Filled subshells contribute to neither $\underline{L}$ nor $\underline{S}$. An atom or ion with more than one valence electron can have a number of spectroscopic terms, e.g. the configuration d^5 generates the 16 terms ^6S, ^4G, ^4F, ^4D, ^4P,

^2I, ^2H, ^2G (twice), ^2F (twice), ^2D (three times), ^2P and ^2S. Hund's rules enable one to know the ground terms: (1) of the terms arising from a given electron configuration, those with the greatest (spin) multiplicity lie lowest; and (2) of the terms with the same multiplicity, the one with the largest value of L lies lowest. With these two rules, it is a simple matter to construct the third column of Table 1-1. There is a third rule which deals with j and is not of concern here for reasons which follow.

When an ion is bound, either in a molecule or in a lattice, the total angular momentum ($\underline{J}$) is not a constant of motion and neither j nor m_j is a useful quantum number. For weak binding, Russell-Saunders coupling still applies more or less quantitatively and, for strong binding, it applies in the sense that spin-spin coupling is stronger than spin-orbit coupling. It is useful here to consider specific cases rather than to generalize. The simplest case of biological interest is that of the cupric ion for which the free ion ground term is $d^9\ ^2D$. Suppose first that the cupric ion is subject to an electric field of octahedral symmetry, and refer to Figure 6-4 for the behavior of the spatial orbitals in this field. Six of the nine d electrons go into the t_{2g} orbitals and the remaining three into the e_g orbitals, leaving a "hole" in the e_g "subshell". If the octahedral field is now tetragonally distorted such that elongation occurs in the z-direction (square planar complex), the $d_{x^2-y^2}$ orbital is left highest in energy and will have single electron occupancy in contrast with all the other orbitals which have two electrons each. The ground state of the ion is equivalent to a filled d-shell (L = 0, S = 0) with a hole (positive charge, L = 2, S = $\frac{1}{2}$) in the $d_{x^2-y^2}$ orbital. More generally, the energy levels of the cupric ion in a square planar complex are as given in Figure 6-5. (Note that these are just the levels of the d-orbitals, but inverted.) The energy differences Δ_i are called "crystal field splittings", having originated in the context of applied electric field effects. The same splittings occur as a result of molecular orbital formation (see below). As compared with the free cupric ion, the bound ion has the same spin but greatly reduced orbital angular momentum. This "quenching" of the orbital angular momentum in bound ions of the iron group has been known experimentally for 50 years and understood theoretically for 40 years. (The reader is referred to Van Vleck, 1932, Chap. XI, for extensive references.) It arises from the removal of the spatial degeneracy of the d-orbitals by reduction of environmental symmetry. Indeed, for any spatially non-degenerate state ("orbital singlet") the expectation values of all components of orbital angular momentum are necessarily zero. There is not, however, a discontinuous change in observables upon an infinitesimal lifting of degeneracies - the behavior depends upon the ratio of off-diagonal matrix elements of the relevant energy operator to the crystal field splitting.

An important case physiologically is that of the ferrous ion, for which the free ion ground term is $d^6\ ^5D$. In an octahedral field of moderate strength the ground electron configuration is $(t_{2g})^4(e_g)^2$ with S = 2. Magnetic susceptibility data indicate that this is the situation in deoxygenated (venous) ferrohemo-

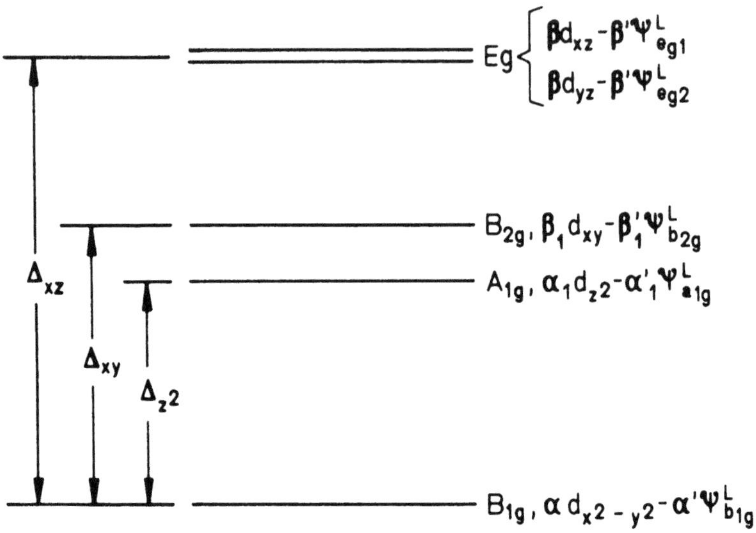

Fig. 6-5. Energy levels of cupric ion in square planar complex

globin. As the strength of the octahedral field increases, the energy required for spin-pairing becomes less than the energy difference between the e_g and t_{2g} orbitals, and the ground configuration changes to $(t_{2g})^6$ with S = 0. The contribution of the iron-oxygen complex to the magnetic susceptibility of an oxygenated hemoglobin solution is consistent with $S_{ground} = 0$ (see also Cerdonio et al., 1976), but one needs to consider here the properties of the O_2 molecule as well. The oxygen molecule in a gas phase and in aqueous solution exhibits a paramagnetic moment corresponding to S = 1 (with no orbital contribution). This situation, unusual among diatomic molecules, is attributed to a two-fold degeneracy in (molecular) orbitals available for two valence electrons. Hund's first rule then requires the total spin to be maximal with the result that the ground state is a triplet. However, when oxygen is bound to hemoglobin, the degeneracy is removed and the spin-paired (singlet) configuration is of lower energy than the triplet. It is a curious feature of the low-spin iron-oxygen complex in oxygenated hemoglobin that a similar quenching of spin angular momentum occurs in both the metal and nonmetal components. The decrease in spin of the ferrous ion in hemoglobin upon oxygenation is one example of the change from a high to a low-spin state which can be induced by complex formation with "strong-field ligands", a phenomenon discussed at various points in this text. The reader should discern that spin quenching is not the same effect as the orbital angular momentum quenching described in the preceding paragraph.

The free ion ground state of the ferric ion is $d^5\ {}^6S$. In an octahedral field of moderate strength, the ground configuration is $(t_{2g})^3(e_g)^2$ which belongs to ${}^6A_1$ (L = 0, S = $\frac{5}{2}$). In a strong octahedral field, the low-spin (S = $\frac{1}{2}$) configuration $(t_{2g})^5$ is the stable one. Introduction of moderate tetragonal and/or rhombic components to the field usually will not alter the spin state.

Spin-orbit coupling is an interaction of the electron spin magnetic moment with the magnetic field associated with the movement of the electron in its orbit through the electric field produced by the nucleus and the other electrons. The corresponding term in the Hamiltonian takes the form

$$H_{S-O} = \lambda \underline{L} \cdot \underline{S}$$

where λ parameterizes the expectation value of a radial operator. $|\lambda|$ is in the range 50 to 1000 cm^{-1} for the iron group ions and is significantly smaller than the usual crystal field splittings which are of the order of 10,000 cm^{-1}. In the case of copper, for example, $\lambda = -828$ cm^{-1} for the free ion and in a cupric complex the separation of the ground state, say $d_{x^2-y^2}$, from the excited d levels is 10,000-20,000 cm^{-1}. If the energy of d_{xy} is Δ_{xy}, then

$$-\frac{\langle d_{xy},\alpha|\lambda\underline{L}\cdot\underline{S}|d_{x^2-y^2},\alpha\rangle}{\Delta_{xy}}|d_{xy},\alpha\rangle = \frac{-i\lambda}{\Delta_{xy}}|d_{xy},\alpha\rangle \approx .05i|d_{xy},\alpha\rangle$$

is mixed into $|d_{x^2-y^2},\alpha\rangle$ by spin-spin coupling and imparts to it a first-order z-component of orbital angular momentum

$$\langle L_z\rangle = -4\lambda/\Delta_{xy}.$$

This quantity is reduced as the ground and excited levels become involved in molecular orbital formation (covalent coordination) as described in the next section. Corresponding to the three components of orbital angular momentum introduced by spin-orbit coupling are components of orbital magnetic moment which, together with the components of spin magnetic moment, determine the interaction of the transition metal ion with an applied magnetic field (Chap. 6-6). The states belonging to an L = 0 free ion term (e.g. those of high-spin Fe(III)) are not connected by spin-orbit interaction.

4. Ligands and Molecular Orbitals

Thus far the discussion has been in terms of metal ions subjected to electrostatic fields of various symmetries. This viewpoint has been useful in the development of the physics of transition metal ions and, when applied to crystals of simple salts, reasonably realistic. However, in biological systems, the metal ions are found in covalent complexes, and models of the structure and function of these complexes are more naturally based upon a framework of chemical bonds.

The atoms, ions, and groups which are bound to the metal ion are called ligands. In metalloproteins, nitrogen is very often one of the atoms bound directly to the metal ion. The ground electron configuration of the free nitrogen atom is $(1s)^2(2s)^2(2p)^3$. In molecules like ammonia (NH_3) and trimethylamine ($N(CH_3)_3$), the five electrons in the 2s and 2p orbitals are distributed among three covalent bonds and a nonbonding "lone pair". As in the familiar tetrahedrally disposed saturated bonds of the carbon atom, the atomic orbitals on the nitrogen atom which are used in this case are four sp^3 hybrids. In three of these are electrons from both the nitrogen and the hydrogen (or carbon) atoms, and in the fourth resides a pair of nitrogen electrons not involved in bonding. The latter pair can, however, be utilized to form "coordinate links", as in trimethylamine oxide, in which both electrons shared between two atoms come from one of the atoms. For example, the hexammonia Werner complexes (e.g. $Cu(II)(NH_3)_6$) have six such links, and the ammonium ion (NH_4^+) can, possibly, be viewed as having one (distributed) coordinate link. Among protein sites, α- and ε-amino groups provide nitrogen atoms with tetrahedral hybrids available for coordination to metal ions (when the pH is high enough for the metal ions to compete successfully with protons for the site).

In the heterocyclic molecule pyridine

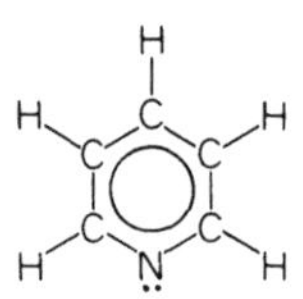

the five valence electrons of the nitrogen are distributed among two localized covalent σ bonds, one nonbonding orbital, and the delocalized π system. The atomic orbitals on the nitrogen which are used for the σ bonds and the lone pair are planar sp^2 trigonal hybrids. The lone pair is available for metal coordination, as in pyridine hemochromogens, but differs somewhat in amount of s character from the lone pairs of amino groups. Of greater biological importance than pyridine are the five-membered heterocyclic pyrrole and imidazole rings of porphyrins and histidine respectively.

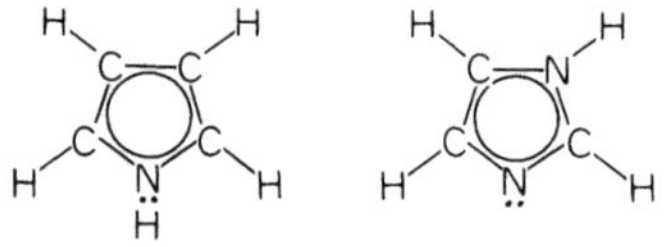

pyrrole imidazole

The pyrrole molecule is shown above in the free acid form. The proton on the nitrogen can be removed in basic solution or displaced by a metal ion (as in heme) to leave a negatively charged group. While the ring has only five atoms, the in-plane bonds about the nitrogen utilize essentially trigonal hybrids. Having received an electron from a hydrogen atom, the nitrogen in pyrrole differs from that in pyridine insofar as it contributes two rather than one electron to the delocalized π system. Imidazole is shown in the alkaline (uncharged) form and is seen to have one pyridine-type nitrogen and one pyrrole-type nitrogen. At acid pH, a proton is bound to the pyridine-type nitrogen. This nitrogen engages in metal coordination when the pH is raised.

The oxygen atom (ground configuration $(1s)^2(2s)^2(2p)^4$), like nitrogen, is often directly bound to metal ions. In "water of coordination" the linkage is through a lone pair on the oxygen. Neutron diffraction-determined structures (of which there are only a limited number) show that the metal-oxygen direction bisects the HOH angle. (A summary of such information is given in Table 6-4 of Hamilton and Ibers, 1968.)

In proteins, ligand oxygens can be provided by ionized carboxyl and phenolic (from tyrosine) groups. Stability is then obtained from ionic bonding in addition to coordinate covalency.

The ground configuration of sulfur, $[Ne](3s)^2(3p)^4$, is related to that of oxygen, but sulfur exhibits differences because of being less electronegative and having d orbitals available for bonding. In the iron-sulfur proteins, both inorganic sulfur and cysteine sulfur are ligands (Chaps. 2-2 h,i, 5-2). In cytochrome c heme iron is coordinated to the sulfur atom of a methionine residue (Chap. 2-2 e) and this may also be the case in peroxidase (Chap. 1-7). Several kinds of evidence point to sulfur as one of the copper ligands in the blue proteins (Chap. 3-3). Along different lines, the blocking of -SH groups on proteins by metal ions is a well-known experimental effect.

The sharing of electrons among a metal ion and its ligands has been inferred from several kinds of measurement and is explicitly demonstrated, for example, by the ligand hyperfine structure seen in EPR spectra. Such systems require a "molecular orbital" description in which the wavefunction is spread over the entire coordination complex. This description is usually in the form of a linear combination of atomic orbitals, one or more of the latter on the metal and one or more on each of the ligands. If we call the combination of orbitals on the ligand ψ_L, and those on the metal ψ_M, these two spatial functions combine in a bonding molecule orbital

$$\psi_b = \alpha\psi_L + \alpha'\psi_M$$

and an antibonding orbital

$$\psi_a = \alpha\psi_M - \alpha'\psi_L$$

where

$$\alpha^2 + (\alpha')^2 \pm 2\alpha\alpha' S = 1,$$

S is the overlap integral $\mathrm{Re}\int \psi_L^* \psi_M d\tau$, and the negative sign in the antibonding orbital indicates that a node of electron density occurs between the metal ion and each ligand. In general, the ligand orbitals are the more stable, and the following diagram applies:

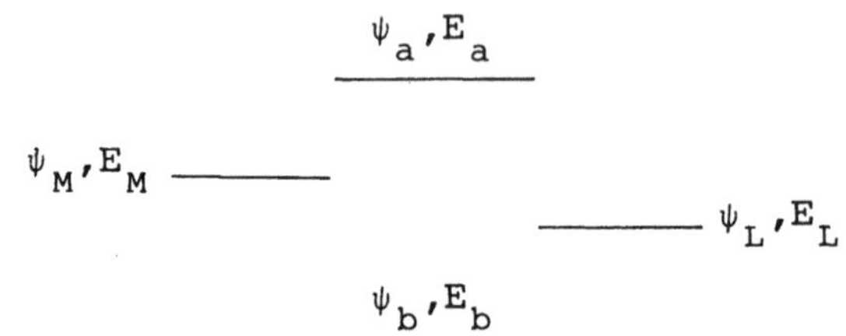

where the bonding orbital ψ_b is mainly ψ_L and the antibonding orbital ψ_a is mainly ψ_M. The energy difference $E_a - E_M$ is the chemical bond equivalent of crystal field splitting, for those ψ_M not engaged in molecular orbital formation remain with energy E_M.

While the metal complexes which occur in proteins are usually of very low symmetry or without any element of symmetry, it is useful to be able to describe these situations in terms of simpler geometries they approximate, namely those of the point groups given in Figure 6-3. The appropriate metal orbitals, ψ_M, are those given earlier and the proper combination of ligand orbitals, ψ_L, to go with a ψ_M is one which transforms in the same way as ψ_M under the symmetry operations of the group. Several examples of ligand combinations which engage in molecular orbital formation with metal d-orbitals are listed in Table 6-2.

When the metal ion is paramagnetic, molecular orbital formation is accompanied by the delocalization of unpaired electron spin onto the ligands. If any of the atoms which share the delocalized spin has a nuclear magnetic moment, both the nuclear and electron paramagnetic resonance spectra will contain information arising from the hyperfine interactions mentioned in Chapter 6-2. Ligand hyperfine effects are treated quantitatively in Chapter 6-7.

The separation of the $d\varepsilon$ from the $d\gamma$ orbitals by electrostatic fields of cubic and lower symmetry was mentioned and illustrated in Chapter 6-1, and, as indicated earlier in the present section, the same effect can result from molecular orbital formation (covalent complexation). Ligands are classed as "strong field" or "weak field" depending upon whether or not their effectiveness in separating the d-orbitals is great enough to alter the spin state of the "free" metal ion. For example, the oxygen molecule, in its diamagnetic complex with ferrous hemoglobin, is a strong field ligand.

Table 6-2. Several combinations of ligand orbitals which engage in molecular orbital formation with metal d-orbitals

Metal Orbital	Ligand orbital combinations[d]					
	Square planar (D_{4h})[a]		Octahedral (O_h)[a]		Tetrahedral (T_d)[b]	
	Combination	Symbol[c]	Combination	Symbol	Combination	Symbol
d_{z^2}	$\frac{1}{2}(\sigma_1 + \sigma_2 + \sigma_3 + \sigma_4)$	a_{1g}	$\frac{1}{\sqrt{12}}(2\sigma_5 + 2\sigma_6 - \sigma_1 - \sigma_2 - \sigma_3 - \sigma_4)$	e_g		
$d_{x^2-y^2}$	$\frac{1}{2}(\sigma_1 - \sigma_2 + \sigma_3 - \sigma_4)$	b_{1g}	$\frac{1}{2}(\sigma_1 - \sigma_2 + \sigma_3 - \sigma_4)$	e_g		
d_{xy}	$\frac{1}{2}(\pi_{y1} + \pi_{x2} - \pi_{y3} - \pi_{x4})$	b_{2g}	$\frac{1}{2}(\pi_{y1} + \pi_{x2} - \pi_{y3} - \pi_{x4})$	t_{2g}	$\frac{1}{2}(\sigma_1 - \sigma_2 - \sigma_3 + \sigma_4)$	t_2
d_{xz}	$\frac{1}{\sqrt{2}}(\pi_{z1} - \pi_{z3})$	e_g	$\frac{1}{2}(\pi_{z1} + \pi_{x5} - \pi_{z3} - \pi_{x6})$	t_{2g}	$\frac{1}{2}(\sigma_1 + \sigma_2 - \sigma_3 - \sigma_4)$	t_2
d_{yz}	$\frac{1}{\sqrt{2}}(\pi_{z2} - \pi_{z4})$	e_g	$\frac{1}{2}(\pi_{z2} + \pi_{y5} - \pi_{z4} - \pi_{y6})$	t_{2g}	$\frac{1}{2}(\sigma_1 - \sigma_2 + \sigma_3 - \sigma_4)$	t_2

[a] Numbering of coordination positions follows the order x, y, -x, -y, z, -z

[b] Numbering of coordination positions follows the order (x,y,z), (-x,y,-z), (x,-y,-z), (-x,-y,z)

[c] Symbol for irreducible representation to which combination belongs

[d] The ligand σ orbitals are taken here to be either spherically symmetric or directed toward the metal ion

5. Absorption of Light

For light to be absorbed by a system of electrons, several conditions must be satisfied. First of all, there must be energy level differences that correspond to the energy of the photons, as expressed in the well-known Bohr frequency condition

$$h\nu \;(\equiv hc/\lambda \equiv hc\bar{\nu}) = \Delta E.$$

The discussion here will be limited to transitions between electronic states (near infra-red, visible, and UV spectra). For such transitions to occur readily in the presence of electromagnetic radiation ("allowed" transitions), the product of the wave functions of the two states must be of odd symmetry with respect to at least one direction in space (say x). That is

$$\psi_m^*(-x)\psi_n(-x) = -\psi_m^*(x)\psi_n(x).$$

Only when this condition is met is it possible for the electric dipole matrix element

$$P_x \equiv X_{mn} \equiv \langle m|ex|n\rangle = e\int \psi_m^* x \psi_n d\tau$$

not to vanish (although it still can). The intensity of an absorption band is quantitated through the experimental oscillator strength

$$f_{mn} = 4.32 \times 10^{-9} \int_{\text{band}} a(\bar{\nu})d\bar{\nu}$$

where $a(\bar{\nu})$ is the molar absorptivity (per cm^{-1}). (It is generally more useful to describe absorptivity in terms of energy rather than wavelength. The absorptivity of a single band as a function of energy is usually symmetric about the peak absorptivity. The individual bands contributing to a solution spectrum are closely Gaussian on a linear energy scale, and the analysis of overlapping bands into components is readily accomplished when the absorptivity is expressed as a function of frequency. A frequency (energy) unit commonly used for the spectral regions in which light absorption occurs for the chromophores discussed here is the kiloKayser (kK) where 1 kK = 1000 cm^{-1}.) The theoretical expression for the oscillator strength arising from an electric dipole transition with matrix elements P_{mn} and energy difference $hc\bar{\nu}_{mn}$ is

$$f_{mn} = 1.085 \times 10^{-5}\bar{\nu}_{mn}\left|\frac{P_{mn}}{e}\right|^2 = 1.085 \times 10^{-5}\bar{\nu}_{mn}|\langle m|\underline{r}|n\rangle|^2$$

where $\bar{\nu}$ is in cm^{-1} and $\underline{r}$ is in Å. Clearly, there is structural information not just in the energy at which maximum absorption occurs, but also in the integrated intensity of the band. Net absorption of light requires the lower energy state to be more densely populated than the one of higher energy, for the quantum-mechanical transition probabilities from high to low and low to high are the same. In the absence of a saturating level of illumination, the Boltzmann distribution insures an excess of population in the lower energy state.

Transitions between quantum states for which all $X_{mn} = 0$ are called "space forbidden", and between quantum states of different spin, "spin forbidden". In real systems which are described to zero order by such states, the transition can still occur but with much lower probability. For example, the product of two atomic orbitals with the same value of ℓ (e.g. $d_{x^2-y^2} \cdot d_{xz}$) is necessarily an even function, and electric dipole transitions between levels so described are space forbidden (Laporte selection rule). Thus, transitions involving rearrangement of the electrons within the valence shell of *free* transition metal ions (spherically symmetric Hamiltonian) are strictly (Laporte) forbidden. However, when such an ion is complexed, there is the possibility that the center of symmetry is lost and an admixture of p-orbitals, for example, can take place. In this case, the zero-order description, which involves only changes among d-orbitals, calls for no absorption of energy, but the process can and does occur by virtue of first and higher order mixing of other orbitals. Vibrations of the metal ion relative to a disposition of ligands which has a center of symmetry can also admix orbitals of differing parity and permit some absorption of light. Thus d-d transitions can occur but are usually weak.

Proteins which have no prosthetic moiety (organic group or metal ion) are colorless (no significant absorption in the visible). There are, however, intense bands in the ultraviolet. Between 300 and 250 nm (33 and 40 kK), the aromatic side chains of the amino acids phenylalanine, tyrosine, and tryptophan have characteristic bands. The tyrosine spectrum changes upon ionization of the phenolic group and is likely to be influenced by metal binding to the phenolic oxygen. Tryptophan does not ionize and is not known to coordinate with metal ions. However, the tryptophan "notch" at 290-295 nm is sensitive to changes in the state of metal ions bound elsewhere. In high-spin ferric hemeproteins, the notch is clearly defined, and loses in resolution as conversion to the low-spin state proceeds. Below 250 nm (40 kK), there are other sets of bands from phenylalanine, tyrosine, and tryptophan, and these are joined by absorbance from histidine, methionine, cystine, cysteine, and from the peptide bond which exhibits maximum absorptivity at 190 nm (53 kK). The side chains which coordinate to metal ions are influenced in their light absorbing characteristics by changes in the state of the metal ion, but these spectral changes are small compared with the overall extremely intense absorption in this region. Considerable care is required if one wishes to utilize the wavelength region below 250 nm to investigate transition-metal ion coordination to proteins (Brill and Sandberg, 1968). In general, the transitions in the amino acid side chains are of two types, π-π between energy levels of the delocalized (molecular) orbitals of the ring systems (e.g. the benzene ring of phenylalanine) and n-π from a non-bonding orbital level to a delocalized orbital.

Porphyrin and flavin absorb in the visible and give characteristic colors to the proteins of which they are prosthetic moieties. The absorption spectra of metalloporphyrins are somewhat simpler than those of the free base porphyrins, the latter being protonated at two of the pyrroles and hence of lower symmetry (two-fold versus four-fold at the center of the molecule). Thus the four visible bands (in the region 500-650 nm) of the free base porphyrins coalesce to two bands called α (at shorter wavelength) and β (longer wavelength), when a metal ion is bound (Platt, 1956). Both the free base and metalloporphyrins have strong absorbance in the near ultra-violet, the Soret band, which peaks at about 400 nm. The Soret and related visible bands arise from π-π transitions, whence the electric dipole transition perpendicular to the molecular plane is not allowed (Chap. 4-4). The in-plane dipole transitions are equivalent in metalloporphyrins, but two visible bands are found as mentioned above. The presence of two bands is attributed to different vibrational levels of the excited electronic state (0-0 and 0-1 vibrational transitions). The latter explanation is supported by the observed constancy of the energy separation between the bands in the spectra of different porphyrins. It should be noted that the 0-1 band is a composite of vibrational components, generally unresolved in absorption spectra because of the small separation in center frequencies but assignable to Raman bands on the basis of their near resonance excitation profiles (Spiro and Strekas, 1972).

When transition-metal ions are present, they will be bound wholly to protein-contributed ligands or to both protein- and organic prosthetic moiety-contributed ligands. The resulting molecular orbital formation will usually influence the position and/or intensity of the ligand transitions (and also the metal transitions but the latter are less likely to be observable). Intense bands, which cannot be directly related either to ligand or to metal transitions, are sometimes found when electron transfer from ligand to metal, or the reverse, can be photochemically induced. Such processes are called "charge-transfer" transitions and require either the presence of an oxidizable (electron donor) ligand (e.g. sulfhydryl group) and a reducible metal [e.g. Cu(II) or Fe(III)], or an oxidizable metal ion [e.g. Cu(I) or Fe(II)] and an electron acceptor ligand. Charge transfer transitions do not differ fundamentally from other electronic transitions; the symmetry and intensity relations given at the beginning of this section apply in the same way to all of these processes.

6. Interaction of Transition-Metal Ions with an Applied Magnetic Field

A system of electrons interacts with an applied magnetic field H according to the following Hamiltonian term, called the Zeeman operator:

$$H_{\text{Zeeman}} = \beta \sum_{\substack{\text{electrons} \\ i}} (\underline{L}_i + 2.0023\,\underline{S}_i) \cdot \underline{H} = \beta(\underline{L} + 2.0023\,\underline{S}) \cdot \underline{H}.$$

β is the Bohr magneton and $\beta\underline{L}$ is the orbital magnetic moment operator, an atomic description entirely predictable from classical electromagnetism. The contribution to the atomic magnetic moment from electron spin, $2.0023\beta\underline{S}$, is essentially twice that expected from the classical relation between magnetic moment and angular momentum. For the magnetic field strengths normally employed in the laboratory (under 30,000 Gauss), Zeeman energies for electron systems are less than 10 cm^{-1}. Thus, the effects of an applied field can be used to study the structure of atoms and molecules without the field perturbing the electronic structure to any significant extent.

The Zeeman interaction energy of the ground state of a bound transition metal ion is calculated by standard perturbation methods. For example, consider the cupric ion in an electrostatic field of square symmetry. The crystal field splittings of the D-manifold are given in Figure 6-5. These energy level differences are of the order of 10,000 cm^{-1} and the spin-orbit energies are of the order of 100 cm^{-1}. One can calculate the spin-orbit perturbed states and then include the Zeeman interaction or, as we shall do below, take the two operators simultaneously.

$$H' = \lambda\underline{L}\cdot\underline{S} + \beta(\underline{L} + 2\underline{S})\cdot\underline{H}.$$

(The factor 2.0023 has been replaced here by 2, for simplicity.) The ground spin state is two-fold degenerate in the absence of an applied field and described as well as is possible by any pair of orthogonal linear combinations (a "Kramers doublet")

$$(\cos\omega)\alpha + (\sin\omega)\beta$$

$$(\sin\omega)\alpha - (\cos\omega)\beta$$

where the trigonometric form of the admixture coefficients is employed for normalization. When the magnetic field is applied, ω is no longer arbitrary but is determined by diagonalization of the matrix of H' within the degenerate manifold (in the usual way prescribed by degenerate perturbation theory.) ω is then found to depend upon the angle which $\underline{H}$ makes with the axis of symmetry (four-fold axis), z. The simplest case is when $\underline{H}$ is parallel to z. Then

$$H' = \lambda\underline{L}\cdot\underline{S} + \beta(L_z + 2S_z)H,$$

$$H'|d_{x^2-y^2};\alpha\rangle = \frac{\lambda}{\sqrt{2}}|d_{xz};\beta\rangle + i\lambda|d_{xy};\alpha\rangle + \beta H(|d_{x^2-y^2};\alpha\rangle + 2i|d_{xy};\alpha\rangle)$$

$$H'|d_{x^2-y^2};\beta\rangle = \frac{\lambda}{\sqrt{2}}|d_{yz};\alpha\rangle - i\lambda|d_{xy};\beta\rangle - \beta H(|d_{x^2-y^2};\beta\rangle - 2i|d_{xy};\beta\rangle)$$

(where, for simplicity, the degenerate states $|d_{xz}\rangle$ and $|d_{yz}\rangle$ have been taken as $|-1\rangle$ and $|1\rangle$ respectively), and the matrix of H' within the spin doublet is

	$\|d_{x^2-y^2};\alpha\rangle$	$\|d_{x^2-y^2};\beta\rangle$
$\langle d_{x^2-y^2};\alpha\|$	βH	0
$\langle d_{x^2-y^2};\beta\|$	0	$-\beta H$.

The matrix is seen to be diagonal as calculated, α and β are themselves correct zero-order spin states, and the first-order contributions to the energy are,

$$\text{for } |+\rangle \equiv |d_{x^2-y^2};\alpha\rangle , \qquad W_+^{(1)} = \beta H,$$

and

$$\text{for } |-\rangle \equiv |d_{x^2-y^2};\beta\rangle , \qquad W_-^{(1)} = -\beta H.$$

The second-order contributions to the energy, obtained from

$$W_k^{(2)} = \sum \frac{|\langle k|\mathcal{H}'|n\rangle|^2}{E_k - E_n},$$

are

$$W_+^{(2)} = \frac{-|\langle d_{xz};\beta|\mathcal{H}'|d_{x^2-y^2};\alpha\rangle|^2}{\Delta_{xz}} - \frac{|\langle d_{xy};\alpha|\mathcal{H}'|d_{x^2-y^2};\alpha\rangle|^2}{\Delta_{xy}}$$

$$= -\frac{\lambda^2}{2\Delta_{xz}} - \frac{(2\beta H + \lambda)^2}{\Delta_{xy}} = -\frac{4\beta H}{\Delta_{xy}} - \frac{4\beta^2 H^2}{\Delta_{xy}} - \lambda^2\left(\frac{1}{\Delta_{xy}} + \frac{1}{2\Delta_{xz}}\right)$$

and

$$W_-^{(2)} = \frac{4\beta H\lambda}{\Delta_{xy}} - \frac{4\beta^2 H^2}{\Delta_{xy}} - \lambda^2\left(\frac{1}{\Delta_{xy}} + \frac{1}{2\Delta_{xz}}\right).$$

Thus, with $\underline{H}$ parallel to the z-axis, the energies of the ground doublet are, to second-order,

$$E_+^{\|} = \beta H\left(1 - \frac{4\lambda}{\Delta_{xy}}\right) - \frac{4\beta^2 H^2}{\Delta_{xy}} - \lambda^2\left(\frac{1}{\Delta_{xy}} + \frac{1}{2\Delta_{xz}}\right)$$

and

$$E_-^{\|} = -\beta H\left(1 - \frac{4\lambda}{\Delta_{xy}}\right) - \frac{4\beta^2 H^2}{\Delta_{xy}} - \lambda^2\left(\frac{1}{\Delta_{xy}} + \frac{1}{2\Delta_{xz}}\right).$$

When $\underline{H}$ is perpendicular to z

$$\mathcal{H} = \lambda\underline{L}\cdot\underline{S} + \beta(L_x + 2S_x)H$$

$$= \lambda\underline{L}\cdot\underline{S} + \beta\left(\frac{L_+ + L_-}{2} + S_+ + S_-\right)H.$$

It can be shown by the method above that $\omega = 45^\circ$, that is

$$|+\rangle = |d_{x^2-y^2};\ (\alpha + \beta)/\sqrt{2}\rangle,$$

and

$$|-\rangle = d_{x^2-y^2};\ (\alpha - \beta)/\sqrt{2}\rangle,$$

and that

$$E_+^\perp = \beta H(1 - \frac{\lambda}{\Delta_{xz}}) - \frac{\beta^2 H^2}{\Delta_{xz}} - \lambda^2(\frac{1}{\Delta_{xy}} + \frac{1}{2\Delta_{xz}}),$$

and

$$E_-^\perp = -\beta H(1 - \frac{\lambda}{\Delta_{xz}}) - \frac{\beta^2 H^2}{\Delta_{xz}} - \lambda^2(\frac{1}{\Delta_{xy}} + \frac{1}{2\Delta_{xz}}).$$

The terms in H^2 and λ^2 are independent of spin state. Thus, while the energy levels (as functions of field strength) exhibit curvature, the difference in energy between spin states is a linear function of H:

$$E_+ - E_- = 2(1 - \frac{a\lambda}{\Delta_i})\beta H$$

where

$$a = 4,\ i = xy \text{ for } H \parallel z \text{ and } a = 1,\ i = xz \text{ for } H \perp z.$$

Electron paramagnetic resonance transitions between the levels E_+ and E_- are allowed and occur in accordance with the Bohr frequency condition $h\nu = E_+ - E_-$. When there is a linear relationship between the strength of the applied magnetic field and the divergence in energy of a spectroscopically accessible pair of levels, it is customary to measure the constant of proportionality in Bohr magnetons, thereby defining the spectroscopic splitting factor or g-value:

$$g \equiv \frac{1}{\beta}\frac{d}{dH}(E_+ - E_-).$$

Then

$$h\nu = g\beta H.$$

For the cupric system above,

$$H \parallel z:\ g_{\parallel} = 2(1 + \frac{4|\lambda|}{\Delta_{xy}}),$$

$$H \perp z:\ g_{\perp} = 2(1 + \frac{|\lambda|}{\Delta_{xz}}).$$

($\lambda \approx 830\ cm^{-1}$ for the cupric ion).

If, instead of square symmetry, the electrostatic field had only two-fold symmetry, $\Delta_{yz} \neq \Delta_{xz}$, and $g_x \neq g_y$.

As mentioned earlier, one can arrive at the g-values by calculating the spin-orbit perturbed ground state and then taking the expectation value of the Zeeman operator for this state. For example

$$g_{||} = 2(\langle L_z \rangle + 2.0023\langle S_z \rangle) = 2(\langle L_z \rangle + 1.0012)$$

$$= 2\left(1 - \frac{4\lambda}{\Delta_{xy}}\right),$$

where the value of $\langle L_z \rangle$ is taken from Chapter 6-3.

The treatment of the cupric ion just given did not include covalency. To a first approximation, delocalization of the cupric hole according to

$$\psi_{x^2-y^2} = \alpha d_{x^2-y^2} - \alpha' \psi_{L,\sigma}$$

$$\psi_{xy} = \beta d_{xy} - \beta' \psi_{L,\pi ip}$$

$$\psi_{xz} = \gamma d_{xz} - \gamma' \psi_{L,\pi op}$$

changes the g-values to

$$g_{||} = 2\left(1 + \frac{4\alpha^2\beta^2|\lambda|}{\Delta_{xy}}\right)$$

and

$$g_{\perp} = 2\left(1 + \frac{\alpha^2\gamma^2|\lambda|}{\Delta_{xz}}\right).$$

(More accurate expressions can be found elsewhere; Kivelson and Nieman, 1961; Kuska and Rogers, 1971.)

One can view the effect of covalency upon the g-values as a reduction of orbital magnetic moment caused by loss of d-character. The ground state $d_{x^2-y^2}$ has been taken above for the cupric ion in square planar symmetry. This is the situation usually found, one which corresponds to stronger sigma bonding in the plane than perpendicular to the plane (elongated octahedral geometry). However compressed octahedral coordination can occur where d_{z^2} is the ground state. The reader is encouraged to carry out the calculations for this case and so arrive at

$$H \parallel z: \; g_{||} = 2$$

$$H \perp z: \; g_{\perp} = 2\left(1 + \frac{3\alpha^2\gamma^2|\lambda|}{\Delta_{xz}}\right)$$

where

$$\psi_{x^2} = \alpha d_{z^2} - \alpha' \psi_{L,\sigma}$$

and

$$\psi_{xz} = \gamma d_{xz} - \gamma' \psi_{L,\pi}.$$

The anisotropy in g for the cupric ion is seen to arise wholly from the orbital contribution. This need not be true in all cases. For example, in high-spin ferric heme the orbital contribution is negligible, and the great anisotropy in g is due to the spin magnetic moment being constrained to a cone which makes a large angle with respect to the heme normal (Fig. 6-6). The latter property is that of the ground doublet, $M_S = \pm \frac{1}{2}$, which is separated from the first excited level, $M_S = \pm \frac{3}{2}$, by a zero-field splitting energy ($2D \gtrsim 10\ cm^{-1}$) which is large compared with photon energies in the frequency range (9 to 35 GHz, to which correspond 0.3 to 1 cm^{-1}) usually employed in electron paramagnetic resonance experiments. The calculation of the g-values proceeds as before except that the ion derives from a ^6S term and the L parts can be dropped from the perturbation operator since they will be without effect:

$$H' = 2\beta \underline{S} \cdot \underline{H}.$$

With H in the xz plane at an arbitrary angle θ from the z-axis,

$$2\beta \underline{S}\cdot\underline{H}|\tfrac{5}{2},\alpha\rangle = \beta H\{\cos\theta\,|\tfrac{5}{2},\alpha\rangle + \sin\theta\,(2\sqrt{2}|\tfrac{5}{2},\tfrac{3}{2}\rangle + 3|\tfrac{5}{2},\beta\rangle)\}$$

and

$$2\beta \underline{S}\cdot\underline{H}|\tfrac{5}{2},\beta\rangle = \beta H\{-\cos\theta\,|\tfrac{5}{2},\beta\rangle + \sin\theta\,(3|\tfrac{5}{2},\alpha\rangle + 2\sqrt{2}\,|\tfrac{5}{2},-\tfrac{3}{2}\rangle)\}.$$

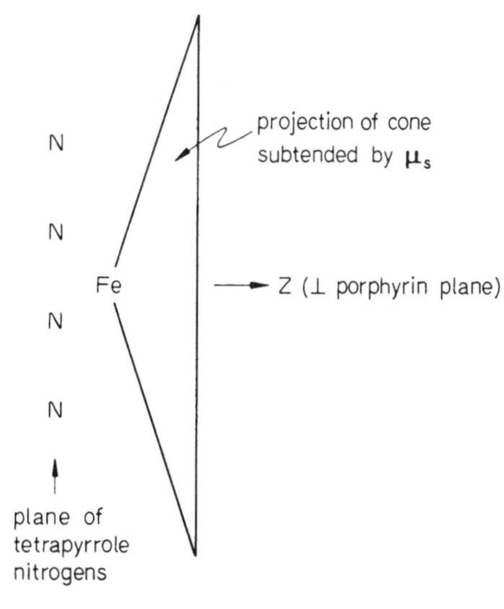

Fig. 6-6. Cone subtended by spin magnetic moment of iron in high-spin ferric heme

The matrix of H' within the ground doublet is then

	$\lvert\alpha\rangle$	$\lvert\beta\rangle$
$\langle\alpha\rvert$	$\beta H \cos\theta$	$3\beta H \sin\theta$
$\langle\beta\rvert$	$3\beta H \sin\theta$	$-\beta H \cos\theta$,

and the energy levels to first order are

$$E_+ = \beta H\sqrt{\cos^2\theta + 9\sin^2\theta} = \frac{\beta H}{2}\sqrt{g_{||}^2\cos^2\theta + g_\perp^2\sin^2\theta} = \frac{g\beta H}{2}$$

and

$$E_- = -\beta H\sqrt{\cos^2\theta + 9\sin^2\theta} = -\frac{\beta H}{2}\sqrt{g_{||}^2\cos^2\theta + g_\perp^2\sin^2\theta} = -\frac{g\beta H}{2}$$

with corresponding states

$$\lvert+\rangle = \frac{1}{\sqrt{2g}}\left\{\frac{g_\perp\sin\theta}{\sqrt{g - g_{||}\cos\theta}}\left\lvert\tfrac{5}{2},\alpha\right\rangle + \sqrt{g - g_{||}\cos\theta}\left\lvert\tfrac{5}{2},\beta\right\rangle\right\}$$

and

$$\lvert-\rangle = \frac{1}{\sqrt{2g}}\left\{\frac{-g_\perp\sin\theta}{\sqrt{g + g_{||}\cos\theta}}\left\lvert\tfrac{5}{2},\alpha\right\rangle + \sqrt{g + g_{||}\cos\theta}\left\lvert\tfrac{5}{2},\beta\right\rangle\right\},$$

where $g_{||} = 2$ ($\underline{H} \parallel z$ which is $\perp$ heme plane), $g_\perp = 6$, and $g = g(\theta) = \sqrt{g_{||}^2\cos^2\theta + g_\perp^2\sin^2\theta} = (E_+ - E_-)/\beta H$. When the symmetry is two-fold or lower, the appearance of the term $E(S_x^2 - S_y^2)$ in the Hamiltonian introduces rhombic character into the calculated g-values. This occurs because $S_x^2 - S_y^2$ mixes the states $M_S = \pm\frac{3}{2}$ and $M_S = \pm\frac{5}{2}$ into the $M_S = \pm\frac{1}{2}$ lowest states. (Scholes (1969) has also considered the effect of spin-orbit coupling with an excited quartet state, the admixture to the ground doublet being of the order of 1%.) To first order $g_{||}$ is unchanged and

$$g_x = 6 - 24E/D,$$

$$g_y = 6 + 24E/D.$$

In the Hamiltonian terms $D(S_z^2 - \frac{1}{3}S(S+1))$ and $E(S_x^2 - S_y^2)$, only spin operators appear, the orbital properties being included in the parameters D and E. Such energy operators differ from ones like e^2/r_{ij} and $\beta(\underline{L} + 2\underline{S})\cdot\underline{H}$ where the orbital operators (distance and momentum) appear explicitly. Hamiltonians can be written in wholly explicit form, wholly in terms of spin-operators, or in mixed form (e.g. $\lambda\underline{L}\cdot\underline{S}$ parameterizes a radial property but leaves the angular momentum operators explicit). In general, knowledge of the spin of a system is the most readily determined information, orbital angular momentum properties more difficult to determine, and distances still more difficult. It is convenient, then, to cast the experimental results in the "spin-Hamiltonian" formalism wherein the electron and nuclear spin operators appear explicitly and the orbital properties are parameterized. Experiments yield a set of such parameters for a given system. The parameters can be quantum mechanically calculated from models. A valid description of the system is among the models which generate parameters consistent with those obtained experimentally.

Derivation of spin-Hamiltonian terms is treated by Abragam and Pryce (1951) and elsewhere (e.g. Bleaney and Stevens, 1953). In general, the method involves an otherwise standard perturbation calculation wherein, however, the spin operators are carried along such that an expression is obtained in which the coefficients (these are the spin-Hamiltonian parameters) of the components of $\underline{S}$ and $\underline{I}$ are functions of the matrix elements of orbital operators. For example, consider the Zeeman and spin-orbit interactions of a system with nondegenerate lowest orbital state $|0\rangle$ and excited states $|n\rangle$.

$$H' = \beta(\underline{L} + 2\underline{S})\cdot\underline{H} + \lambda\underline{L}\cdot\underline{S}$$

$$\langle 0|H'|0\rangle = \beta\langle 0|\underline{L}|0\rangle\cdot\underline{H} + 2\underline{S}\cdot\underline{H}\langle 0|0\rangle + \lambda\langle 0|\underline{L}|0\rangle\cdot\langle 0|\underline{S}|0\rangle.$$

It can be shown (van Vleck, 1932) that the expectation values of all three components of the orbital angular momentum are zero for nondegenerate orbitals. Thus,

$$\langle 0|H'|0\rangle = 2\underline{S}\cdot\underline{H}.$$

Proceeding to the second order perturbation calculation

$$-\sum_{n\neq 0}\frac{|\langle 0|H'|n\rangle|^2}{E(n) - E(o)}$$

we find that the term of H' which does not contain orbital operators is zero because $\langle 0|n\rangle$ is zero, and that the other two terms lead to a form quadratic in the components of $\underline{S}$ and $\underline{H}$:

$$-\beta^2\Lambda_{ij}H_iH_j - \beta f_{ij}H_iS_j + d_{ij}S_iS_j$$

where $i,j = x,y,z$ and Λ_{ij}, f_{ij}, and d_{ij} are functions of orbital angular momentum matrix elements between the ground and excited states. Collecting the first and second order contributions, we have the spin-Hamiltonian

$$H_s = \sum_{i,j}\{\beta(2\delta_{ij} - f_{ij})H_iS_j - \beta^2\Lambda_{ij}H_iH_j + d_{ij}S_iS_j\}$$

where δ_{ij} is the Kronecker delta. The second term expresses temperature independent paramagnetism and the third term, zero field splitting. In the appropriate coordinate system (principal axes of the tensor f_{ij}), the first term takes the form

$$\beta(g_zH_zS_z + g_xH_xS_x + g_yH_yS_y)$$

where

$$g_i = 2\left(1 - \lambda \sum_{n \neq o} \frac{|\langle o|L_i|n\rangle|^2}{E(n) - E(o)}\right).$$

The reader should demonstrate for himself that the latter formula leads to the g-values given above for the cupric ion. Provided that the same coordinate system simplifies all of its terms, the complete spin-Hamiltonian is

$$H_s = \beta(g_z H_z S_z + g_x H_x S_x + g_y H_y S_y) + D\{S_z^2 - \tfrac{1}{3}S(S+1)\} + E(S_x^2 - S_y^2)$$

$$+ A_z S_z I_z + A_x S_x I_x + A_y S_y I_y + Q\{I_z^2 - \tfrac{1}{3}I(I+1) + \eta(I_x^2 - I_y^2)\}$$

$$- g_n \beta_n \underline{H} \cdot \underline{I},$$

exclusive of ligand interactions. The final term is the direct interaction of the nuclear moment with the applied field, of relatively small energy.

In some cases, special consideration must be given to the relation between the spin-Hamiltonian g's and the spectroscopic splitting factors (experimental g-values). In the case of the high-spin ferric ion in heme, for example, the measured g-values of 2 and 6 can be calculated on first principles as was done above. The large g-value cannot arise from an orbital contribution because there is none for any orientation of $\underline{H}$. Instead it arises from the fact that the zero-field splitting is many times greater than the Zeeman and microwave photon energies, and only transitions within $M_s = \pm\frac{1}{2}$ are observed. One can, therefore, alternatively describe the system in terms of a fictitious doublet ($S' = \frac{1}{2}$) with spin-Hamiltonian

$$H_s' = \beta(g_z H_z S_z' + g_x H_x S_x' + g_y H_y S_y')$$

where the g_i are the experimental values and the prime refers the spin operator to the representation $|\frac{1}{2},\alpha\rangle, |\frac{1}{2},\beta\rangle$. While the proper model for this system and a lower limit on D were suggested on the basis of electron paramagnetic resonance data alone (Bennett et al., 1955; George et al., 1956) other experimental methods were required to measure D-values and thereby to establish the model (Feher and Richards, 1967; Kotani and Morimoto, 1967; Scholes et al., 1971; Tasaki et al., 1971). (Often, however, zero-field splittings are smaller than the microwave photon energies and can be measured by EPR.)

The interactions of transition metal ions with a magnetic field are most often studied by means of the magnetic susceptibility measurement, electron and nuclear magnetic resonance, and Mössbauer spectroscopy. The reader is referred to recent reviews of these methods as applied to biological systems (e.g. Brill, 1969; Franconi, 1971; Ingram, 1969; Lang, 1970; Phillips, 1971; Swartz et al., 1972). Figure 6-7 illustrates how spin-Hamiltonian para-

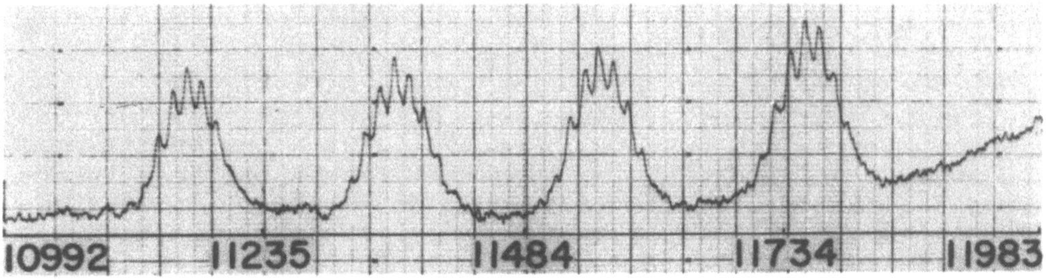

Fig. 6-7. Low field EPR spectrum of ^{63}Cu protoporphyrin IX dimethyl ester diluted in metal-free porphyrin. ν = 35.197 GHz (Hsu, 1971)

meters are obtained from electron paramagnetic resonance spectra. The spectrum shown is from cupric dimethyl protoporphyrin IX diluted in the metal-free porphyrin (Hsu, 1971). The first derivative of the energy absorbed with respect to the magnetic field, that is dA/dH, is recorded as a function of H. From polycrystalline samples of axial complexes, the latter display produces absorption-like bands at the unique (z or $\parallel$) g-value and derivative-like bands at the equal (x, y or $\perp$) g-values. The copper nucleus has a spin I of $\frac{3}{2}$, so that each resonance is split into $2I + 1 = 4$ hyperfine bands. Furthermore, there are four covalently bonded, essentially equivalent, nitrogens, each with a nuclear spin of 1, so that each copper hyperfine band is split into $2(4) + 1 = 9$ ligand hyperfine lines. In the low-field ($g_{\parallel}$) resonance, all of these effects are shown. The central field (H_c) gives $g_{\parallel}$ from the formula

$$g = h\nu/\beta H = 0.71449\nu(GH_z)/H_c(KG)$$

$$= (0.71449)(35.197)/(11.473) = 2.192.$$

The value of $A_{\parallel}$ is obtained from the splittings $\Delta H_{\parallel}^{Cu}$ according to

$$A(\text{in } cm^{-1}) = g\beta\Delta H = g_{\parallel}\,(4.6686 \times 10^{-5})\Delta H_{\parallel}^{Cu}\ (\text{in Gauss})$$

$$= (2.192)(4.6686 \times 10^{-5})(199) = 0.0204\ cm^{-1}.$$

The value of the ligand hyperfine splitting parameter $A_{\perp}^{N}$ (where $\perp$ in this connection means perpendicular to the Cu-N bond) is obtained from the ligand splitting $\Delta H_{\perp}^{N}$ according to

$$A = g_{\parallel}\,\beta\Delta H_{\perp}^{N} = (2.192)(4.6686 \times 10^{-5})(14.1) = 0.00144\ cm^{-1}.$$

The following section quantitatively treats ligand hyperfine splitting and related effects.

7. Magnetic Interactions of the Metal Electrons with Nuclei in the Coordination Sphere

There can be significant interactions of the electronic orbital and spin moments of a transition metal ion with the nuclear magnetic moments of nearby atoms. When the effects of these interactions are measurable, the data leads to estimates of covalency (delocalization) and/or distance of separation. In the best circumstances, information about hybridization can also be available

Suppose, for example, that the metal orbital ψ_M, containing an unpaired spin, is engaged in a molecular orbital with a pyridine nitrogen through a trigonal hybrid ψ_N:

$$\psi = \alpha\psi_M - \alpha_N\psi_N - \sum_{\substack{\text{other}\\ \text{ligands}}} \alpha_L\psi_L .$$

The amount of s-character in ψ at the nitrogen is $(\alpha_N)^2/3$ and the coefficient of the Fermi part of the ligand hyperfine interaction ($A_F^N \underline{I}_N \cdot \underline{S}$) is given by

$$A_F^N = \frac{(\alpha_N)^2}{3} \frac{1}{2S} \frac{16}{3} \pi g_N \beta\beta_N |2s^N(0)|^2$$

where $2s^N(0)$ is the value of the nitrogen 2s wave function at the nucleus.[6] (The S in the denominator is the electronic spin quantum number, e.g. $\frac{5}{2}$ for high-spin Fe(III).) (We assume here that core polarization of the nitrogen 1s closed shell from unpaired spin in the 2s and 2p orbitals is negligible.) $A_F^N \underline{I}_N \cdot \underline{S}$ gives rise to an isotropic (apart from the effect of possible anisotropy in the electronic magnetic moment) splitting of the electron paramagnetic resonances into three lines ($2I_N + 1 = 3$). For a typical delocalization $(\alpha_N)^2$ of 8%, $A_F^N = 0.0014\ \text{cm}^{-1}/2S$; the corresponding splitting in Gauss is, apart from the effect of $\underline{I}_N \cdot \underline{S}$[7], $A_F^N/g\beta = 15.0/Sg$. It is often possible to obtain EPR spectra under conditions where nitrogen hyperfine splittings are resolvable, as in Figure 3-7. One has also to consider the p-character in ψ at the nitrogen, which for a trigonal hybrid is $\frac{2}{3}(\alpha_N)^2$. This p-orbital admixing, through what can be called a "covalent dipole-dipole" interaction (the orbital field contribution is, however, included), contributes a relatively small anisotropic component to the ligand hyperfine splitting. It can be shown that the ligand hyperfine interaction parameter for an unpaired spin in a nitrogen trigonal hybrid has a value $A_{\parallel}^N$ in the bond direction 12.4% greater than the Fermi coefficient, and a value $A_{\perp}^N$ perpendicular to the bond direction which is 6.2% less than A_F^N. The EPR data can be analyzed to obtain A_F^N which yields, in terms of sp^h hybridized atomic orbitals

---

[6] The value of $|2s^N(o)|^2$ to be used is the average of the Hartree (1957) and Clementi (1964) values

[7] In high-spin ferric heme, $\underline{I}_N \cdot \underline{S}$ introduces a large orientation dependence (Scholes, 1970)

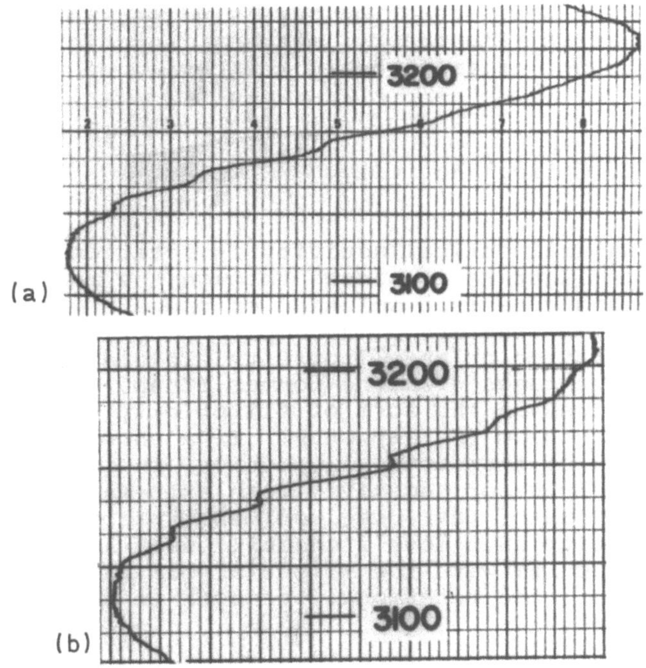

Fig. 6-8. (a) Electron paramagnetic resonance spectrum from a single crystal of cupric insulin prepared in ordinary water buffer. The spectrum was taken at 112 K, in the ab plane at $\phi = 16^{\circ}$, modulation amplitude 1.5 G peak-to-peak, microwave frequency 9.141 GHz. (b) Electron paramagnetic resonance spectrum from a single crystal of cupric insulin equilibrated against heavy water buffer for 2 days. The spectrum was taken at 115 K, in the ab plane at $\phi = 22^{\circ}$, modulation amplitude 1.0 G peak-to-peak, microwave frequency 9.142 GHz

$$(\alpha_N)^2 = 3S(h + 1)A_F^N/8\pi g_N \beta\beta_N |2s^N(0)|^2.$$

This formula for delocalization is as accurate as the estimate of hybridization (h), all other factors being more accurately known. Application of the formula to the value of A_F^N obtained from Figure 6-7 (A_F^N = 1.07 $A_\perp^N$ = 0.00154 cm^{-1}) gives $(\alpha_N)^2 = 0.0875$ for h = 2. There are four essentially equivalent pyrrole nitrogens coordinated to the cupric ion in this porphyrin complex, so that the total delocalization of the $d_{x^2-y^2}$ cupric hole is 0.34. The reader is referred to Owen and Thornley (1966) for a broader treatment of ligand hyperfine interactions and some critical discussion of the approximations involved.

There are noncovalent dipole-dipole interactions between a paramagnetic metal ion and the nuclear magnetic moments of atoms bound to it, and also of atoms further away. The former interactions contribute to the ligand hyperfine parameters discussed in the preceding paragraph. The latter interactions are usually not resolvable and contribute to the linewidth. The effect on the EPR spectrum (at specific orientations of the magnetic field) of substituting D_2O for H_2O in cupric insulin crystals is shown in Figure 6-8 (Brill and Venable, 1972). The nuclear magnetic moment of the deuteron is about one-third that of the proton. The resonance line narrowing, together with other data, indicates that water molecules occupy three of the six coordination positions about each metal ion. The measured decrease in linewidth can be used to calculate the average proton-to-metal distance. Average nuclei-to-metal distances for more distant nuclei can also be obtained from NMR relaxation rate studies. For an exposition of NMR relaxation rate theory, methods, and biological applications, the reader is referred to the review of Mildvan and Cohn (1970).

8. Optical Activity

Natural optical activity is rotation of the plane of polarization of light as it passes through the medium of interest, which in this case is a solution of protein molecules to which transition metal ions are bound. Optical activity can be measured as optical rotation or as ellipticity. The latter measurement has certain advantages over the former (Chap. 1-6) and most data is now obtained in the form of ellipticity versus wavelengths close to the dichroism (CD) spectrum. If a chromophore has several transitions, all of them optically active, the CD spectrum consists of bands, positive or negative, with extrema at wavelengths close to the maxima of the (resolved) absorption bands. The molecular ellipticity unit most frequently used is deg cm^2/decimol. From the CD spectrum with the molecular ellipticity $[\theta]$ expressed in this unit, the rotational strength of the band at $\bar{\nu}_{mn}$ is quantitated through

$$R_{mn} = 6.95 \times 10^{-43} \int_{\text{band}} \frac{[\theta]}{\bar{\nu}} d\bar{\nu} .$$

The theoretical expression for the rotational strength arising from a transition with electric dipole matrix element $\underline{P}_{mn}$ and magnetic dipole matrix element $\underline{M}_{mn} \equiv \langle m|\beta\underline{L}|n\rangle$ is

$$R_{mn} = \text{imaginary part of } (\underline{P}_{mn} \cdot \underline{M}_{nm})$$

$$= 4.454 \times 10^{-38} \times \text{imaginary part of } \langle m|\underline{r}|n\rangle \cdot \langle n|\underline{L}|m\rangle$$

where $\underline{r}$ is in Å and $\underline{L}$ in units of $\hbar$.

In a general discussion of the optical activity of groups of approximate symmetry, one- and two (or more)-electron theories are considered. Two-electron processes involving electric dipole-electric dipole coupled oscillators cannot be effective for d-d transitions because these are fundamentally weak. A two-electron process which can apply here is the interaction between a ligand chromophoric group and the metal ion such that a magnetically-allowed d-d transition (these are indicated in Table 6-1) is coupled to an electrically allowed ligand transition. The effectiveness of such a mechanism for "borrowing" intensity depends upon the energy of the ligand transition not differing greatly from that of the d-d transition, a situation which does not apply for amino acid side-chains but can for organic prosthetic groups. When operative, as in heme, this case can present difficulty in the analysis of the data for the intense ligand bands can obscure the metal absorption.

The one-electron processes depend upon the environment of the group being dissymmetric and introducing electric-dipole transition moment, either through hybridization of the metal orbitals or through admixture of charge-transfer states into the ion states. There is some analogy here to the case of coupling to a ligand transition and, similarly the charge-transfer transition cannot differ greatly in energy from the d-d transition for it to contribute significantly to the electric moment. When more than the expected number of bands is found in the region of d-d transitions, and no ligand absorption can be assigned to this region, then charge-transfer transitions are implicated. This is likely to be the case for the intense cupric chromophore in blue proteins. In this situation some of the d-d bands may exhibit heightened absorbance arising from coupling of ion states to charge-transfer states.

The reader is referred to Schellman (1966) for a systematic treatment of the symmetry requirements for perturbing potentials to produce optical rotation in nonactive one-electron states, and to Richardson (1971 and later papers) for illuminating development and applications of the theory to transition metal compounds. For complexes of geometries which deviate little from ideal point group symmetries there are "sector rules" which relate the sign of the net circular dichroism to certain structural features (Martin, 1974). In the case of metal binding to protein groups, the departure from ideal symmetry may often be so great as to render inapplicable the use of simple sector rules.

Hybridization of the d-orbitals with small amounts of p-functions produces first order quantum mechanical models of transition metal ions from which oscillator and rotational strengths can be calculated and compared with the experimental values. (The model must, of course, be consistent with the spin-Hamiltonian parameters measured by magnetic methods.) The radial matrix elements which appear in the oscillator and rotational strengths are theoretical parameters which are part of the characterization of the model. If these have the length appropriate to metal ion wavefunctions, no charge-transfer is implied. If the radial matrix elements are required to be larger than atomic, a charge-transfer state is implicated.

References

Aasa, R., Vänngård, T.: EPR signal intensity and powder shapes: a reexamination. J. Mag. Res. 19, 308-315 (1975)

Abragam, A., Pryce, M.H.L.: Theory of the nuclear hyperfine structure of paramagnetic resonance spectra in crystals. Proc. Roy. Soc. (London), Ser. A 205, 135-153 (1951)

Ackermann, E., Brill, A.S.: Magnetic and spectrophotometric studies of the kinetics of the catalysis of xanthine oxidation by xanthine oxidase from cow's milk. Biochim. Biophys. Acta 56, 397-412 (1962)

Adams, A., Klemperer, W., Dunn, T.M.: Rotational analysis of the blue-green system of scandium oxide. Can. J. Phys. 46, 2213-2220 (1968)

Adams, M.J., Blundell, T.L., Dodson, E.J., Dodson, G.G., Vijayan, M., Baker, E.N., Harding, M.M., Hodgkin, D.C., Rimmer, B., Sheats, S.: Structure of rhombohedral 2 zinc insulin crystals. Nature (London) 224, 491-495 (1969)

Adman, E.T., Sieker, L.C., Jensen, L.H.: The structure of a bacterial ferredoxin. J. Biol. Chem. 248, 3987-3966 (1973)

Agrawal, B.B.L., Margoliash, E., Levenberg, M.I., Egan, R.S., Studier, M.H.: Identification of product of reaction of 3-amino-1,2,4-triazole with catalase-H_2O_2 complex I. Federation Proc. 29, 732 (1970)

Alberding, N., Austin, R.H., Beeson, K.W., Chan, S.S., Eisenstein, L., Frauenfelder, H., Nordlund, T.M.: Tunneling in ligand binding to heme proteins. Science 192, 1002-1004 (1976)

Alberding, N., Austin, R.H., Chan, S.S., Eisenstein, L., Frauenfelder, H., Gunsalus, I.C., Nordlund, T.M.: Dynamics of carbon monoxide binding to protoheme. J. Chem. Phys. 65, 4701-4711 (1976)

Alpert, Y., Couder, Y., Tuchendler, J., Thomé, H.: Determination of the zero-field splitting in human acid methemoglobin by millimeter and submillimeter ESR experiments. Biochim. Biophys. Acta 322, 34-37 (1973)

Ambler, R.P.: Species differences in the amino acid sequences of bacterial proteins. In: Chemotaxonomy and Serotaxonomy, pp. 57-64. Hawkes, J.G. (ed.). London, New York: Academic Press 1968

Ambler, R.P., Brown, L.H.: The amino acid sequence of *Pseudomonas fluorescens* azurin. Biochem. J. 104, 784-825 (1967)

Andréasson, L.-E., Vänngård, T.: Evidence of a specific copper (II) in human ceruloplasmin as a binding site for inhibitory anions. Biochim. Biophys. Acta 200, 247-257 (1970)

Angelici, R.J.: Stability of Coordination Compounds. Inorganic Biochemistry, Vol. 1, Chap. 2. Eichhorn, G.L. (ed.). Amsterdam, London, New York: Elsevier 1973

Antanaitis, B.C., Moss, T.H.: Magnetic studies of the four-iron high-potential, non-heme protein from *Chromatium Vinosum*. Biochim. Biophys. Acta 405, 262-279 (1975)

Antonini, A., Brunori, M.: Hemoglobin and Myoglobin in Their Reactions with Ligands. Amsterdam, London: North-Holland 1971

Asakura, T., Yonetani, T.: Studies on cytochrome c peroxidase. XVIII. Recombination of apoenzyme with protoporphyrin and protoheme monomethyl esters. J. Biol. Chem. 247, 2278-2282 (1972)

Assour, J.M.: Electron spin resonance of tetraphenylporphine chelates. J. Chem. Phys. 43, 2477-2489 (1965)
Austin, R.H., Beeson, K.W., Eisenstein, L., Frauenfelder, H., Gunsalus, I.C.: Dynamics of ligand binding to myoglobin. Biochemistry 16, 5355-5373 (1975)
Bates, C.A., Moore, W.S., Standley, K.J., Stevens, K.W.H.: Paramagnetic resonance of a Cu^{2+} ion in a tetrahedral crystal field. Proc. Phys. Soc. 79, 73-83 (1962)
Bearden, A.J., Dunham, W.R.: Iron electronic configurations in proteins: Studies by Mössbauer spectroscopy. Struct. Bonding 8, 1-52 (1970)
Beinert, H.: Flavins and flavoproteins, including iron-sulfur proteins. In: Biological Applications of Electron Spin Resonance, Chap. 8. Swartz, H.M., Bolton, J.R., Borg, D.C. (eds.). New York, London: Wiley-Interscience 1972
Beinert, H., Griffiths, D., Wharton, D., Sands, R.: Properties of the copper associated with cytochrome oxidase as studied by paramagnetic resonance spectroscopy. J. Biol. Chem. 237, 2337-2346 (1962)
Beinert, H., Sands, R.H.: Studies on succinic and DPNH dehydrogenase preparations by paramagnetic resonance (EPR) spectroscopy. Biochem. Biophys. Res. Commun. 3, 41-46 (1960)
Bemski, G., Nagel, R.L.: Electron spin resonance of four human hemoglobins. Biochim. Biophys. Acta 154, 592-595 (1968)
Bennett, J.E., Ingram, D.J.E., George, P., Griffith, J.S.: Paramagnetic resonance absorption of ferrihaemoglobin and ferrimyoglobin derivatives. Nature (London) 176, 394 (1955)
Bentley, G., Dodson, E., Dodson, G., Hodgkin, D., Mercola, D.: Structure of insulin in 4-zinc insulin. Nature (London) 261, 166-168 (1976)
Bleaney, B., Stevens, K.W.H.: Paramagnetic resonance. Rep. Progr. Phys. 16, 108-159 (1953)
Blumberg, W.E.: Some aspects of models of copper complexes. In: The Biochemistry of Copper, pp. 49-64. Peisach, J., Aisen, P., Blumberg, W.E. (eds.). New York, London: Academic Press 1966
Blumberg, W.E.: The EPR of high spin Fe^{3+} in rhombic fields. In: Magnetic Resonance in Biological Systems, pp. 119-133. Ehrenberg, A., Malmström, B.G., Vänngård, T. (eds.). Oxford, London: Pergamon 1967
Blumberg, W.E., Eisinger, J., Aisen, P., Morell, A.G., Scheinberg, I.H.: Physical and chemical studies on ceruloplasmin. I. The relation between blue color and the valence states of the copper. J. Biol. Chem. 238, 1675-1682 (1963)
Blumberg, W.E., Goldstein, M., Lauber, E., Peisach, J.: Magnetic resonance studies on the mechanism of the enzymic β-hydroxylation of 3,4-dihydroxyphenylethylamine. Biochim. Biophys. Acta 99, 187-190 (1965)
Blumberg, W.E., Horecker, B., Kelly-Falcoz, F., Peisach, J.: The role of copper in galactose oxidase. Biochim. Biophys. Acta 96, 336-338 (1965)
Blumberg, W.E., Peisach, J.: The optical and magnetic properties of copper in *Chenopodium Album* plastocyanin. Biochim. Biophys. Acta 126, 269-273 (1966)
Blundell, T.L., Cutfield, J.F., Cutfield, S.M., Dodson, E.J., Dodson, G.G., Hodgkin, D.C., Mercola, D.A., Vijayan, M.: Atomic positions in rhombohedral 2-zinc insulin crystals. Nature (London) 231, 506-511 (1971)
Boas, J.F., Pilbrow, J.R., Hartzell, C.R., Smith, T.D.: ESR studies of some copper (II) peptide complexes. J. Chem. Soc. (A), 572-577 (1969)
Boeri, E., Ehrenberg, A., Paul, K.G., Theorell, H.: On the compounds of ferricytochrome c appearing in acid solution. Biochim. Biophys. Acta 12, 273-282 (1953)
Bolton, W., Perutz, M.F.: Three-dimensional Fourier synthesis of horse deoxyhaemoglobin at 2.8 Å resolution. Nature (London) 228, 551-556 (1970)

Bowers, K.D., Owen, J.: Paramagnetic resonance: II. Rep. Prog. Phys. 18, 304-373 (1955)
Boyer, P.D. (ed.): The Enzymes, 3rd. ed., Vols. XI, XII, XIII. New York: Academic Press 1975, 1976
Brackett, G.C., Richards, P.L., Caughey, W.S.: Far-infrared magnetic resonance in Fe (III) and Mn (III) porphyrins, myoglobin, hemoglobin, ferrichrome A, and Fe (III) dithio-carbamates. J. Chem. Phys. 54, 4383-4401 (1971)
Braterman, P.S., Davies, R.C., Williams, R.J.P.: The properties of metal-porphyrins and similar complexes. Advanc. Chem. Phys. 7, 359-407 (1964)
Bray, R.C.: Sudden freezing as a technique for the study of rapid reactions. Biochem. J. 81, 189-193 (1961)
Bray, R.C.: Molybdenum iron-sulfur flavin hydroxylases and related enzymes. In: The Enzymes, Vol. XII, Chap. 6. Boyer, P.D. (ed.). New York: Academic Press 1975
Bray, R.C., Knowles, P.F., Meriwether, L.S.: ESR and the role of molybdenum in enzymic catalysis by milk xanthine oxidase. In: Magnetic Resonances in Biological Systems, pp. 249-264. Ehrenberg, A., Malmström, B.G., Vänngård, T. (eds.). Oxford, London: Pergamon 1967
Bray, R.C., Malmström, B.G., Vänngård, T.: The chemistry of xanthine oxidase. 5. Electron-spin resonance of xanthine oxidase solutions. Biochem. J. 73, 193-197 (1959)
Bray, R.C., Meriwether, L.S.: Electron spin resonance of xanthine oxidase substituted with molybdenum-95. Nature (London) 212, 467-469 (1966)
Bray, R.C., Pettersson, R.: Electron-spin-resonance measurements. Biochem. J. 81, 194-195 (1961)
Bray, R.C., Pettersson, R., Ehrenberg, A.: The chemistry of xanthine oxidase. 7. The anaerobic reduction of xanthine oxidase studied by electron-spin resonance and magnetic susceptibility. Biochem. J. 81, 178-189 (1961)
Bray, R.C., Swann, J.C.: Molybdenum-containing enzymes. Struct. Bonding 11, 107-144 (1972)
Brill, A.S.: Peroxidases and catalase. In: Comprehensive Biochemistry. Vol. XIV, Chap. 10. Florkin, M., Stotz, E.H. (eds.). Amsterdam, London, New York: Elsevier 1966
Brill, A.S.: Magnetic susceptibility. In: Physical Techniques in Biological Research, Vol. IIB, pp. 136-147, 213-216. Moore, D. (ed.). New York: Academic Press, 1969
Brill, A.S.: Iron out-of-planarity and tetrapyrrole nitrogen nuclear spin state mixing in high-spin ferric heme. Molec. Phys. 24, 787-800 (1972)
Brill, A.S., Bryce, G.F.: Cupric ion in blue proteins. J. Chem. Phys. 48, 4398-4404 (1968)
Brill, A.S., Bryce, G.F., Maria, H.J.: Optical and magnetic properties of *Pseudomonas* azurins. Biochim. Biophys. Acta 154, 342-351 (1968)
Brill, A.S., Kirkpatrick, P.R., Scholes, C.P., Venable, J.H., Jr.: Electron paramagnetic resonance of single crystals as an environmental probe. In: Probes of Structure and Function of Macromolecules and Membranes. Vol. I, pp. 135-141. Chance, B.C., Lee, C.-P., Blasie, J.K. (eds.). New York, London: Academic Press 1971
Brill, A.S., Martin, R.B., Williams, R.J.P.: Copper in biological systems. In: Electronic Aspects of Biochemistry, pp. 519-557. Pullman, B. (ed.). New York: Academic Press 1964
Brill, A.S., Sandberg, H.E.: Spin-state-dependent hemoprotein ultraviolet-absorption bands. Proc. Natl. Acad. Sci. 57, 136-140 (1967)
Brill, A.S., Sandberg, H.E.: Spectral studies of iron coordination in hemeprotein complexes: Difference spectroscopy below 250 mμ. Biophys. J. 8, 669-690 (1968)

Brill, A.S., Shyr, C.-I., Walker, T.C.: Power saturation of electron paramagnetic resonances from high-spin ferric haemproteins at 4.2 K. Molec. Phys. 29, 437-454 (1975)

Brill, A.S., Venable, J.H., Jr.: The binding of transition metal ions in insulin crystals. J. Molec. Biol. 36, 343-353 (1968)

Brill, A.S., Venable, J.H., Jr.: Water of coordination in insulin. J. Molec. Biol. 66, 169-180 (1972)

Brill, A.S., Weinryb, I.: Reactions of horseradish peroxidase with azide. Evidence for a methionine residue at the active site. Biochemistry 6, 3528-3535 (1967)

Brill, A.S., Williams, R.J.P.: The absorption spectra, magnetic moments, and the binding of iron in some haemoproteins. Biochem. J. 78, 246-253 (1961)

Bryce, G.F.: Electron paramagnetic resonance study of cupric-peptide complexes. J. Phys. Chem. 70, 3549-3557 (1966)

Bryce, G.F., Gurd, F.R.N.: Visible spectra and optical rotatory properties of cupric ion complexes of L-histidine-containing peptides. J. Biol. Chem. 241, 122-129 (1966)

Buchanan, B.B., Arnon, D.I.: Ferredoxins: chemistry and function in photosynthesis, nitrogen fixation, and fermentative metabolism. Advan. Enzymol. 33, 119-176 (1970)

Buckingham, D.A.: Structure and Stereochemistry of Coordination Compounds. Inorganic Biochemistry, Vol. I, Chap. 1. Eichhorn, G.L. (ed.). Amsterdam, London, New York: Elsevier 1973

Burns, R.C., Hardy, R.W.F.: Nitrogen Fixation in Bacteria and Higher Plants. Berlin-Heidelberg-New York: Springer 1975

Campbell, I.D., Dobson, C.M., Williams, R.J.P.: Assignment of the ^{1}H N.M.R. spectra of proteins. Proc. Roy. Soc. (London), Ser. A 345, 23-40 (1975)

Cantarow, A., Schepartz, B.: Biochemistry. Philadephia, London: Saunders 1967

Carrell, H.L., Glusker, J.P., Job, R., Bruice, T.C.: A synthetic tetranuclear iron-sulfur complex with ionized side chains: the crystal structure of $(Fe_4S_4(S(CH_2)_2COO)_4)^{6-} \cdot Na_5 \cdot N(C_4H_9)_4{}^{6+} \cdot 5\ C_5H_9NO$. J. Am. Chem. Soc. 99, 3683-3690 (1977)

Carrico, R.J., Malmström, B.G., Vänngård, T.: Anaerobic oxidation-reduction titrations of human ceruloplasmin. Evidence for diamagnetic electron acceptors in the protein. Europ. J. Biochem. 20, 518-524 (1971)

Carter, C.W., Jr., Freer, S.T., Xuong, NG.H., Alden, R.A., Kraut, J.: Structure of iron-sulfur cluster in the *Chromatium* iron protein at 2.25 Å resolution. Cold Spr. Harbor Symp. Quant. Biol. 36, 381-385 (1971)

Carter, C.W., Jr., Kraut, J., Freer, S.T., Alden, R.A.: Comparison of oxidation-reduction site geometries in oxidized and reduced *Chromatium* high potential iron protein and oxidized *Peptococcus aerogenes* ferredoxin. J. Biol. Chem. 249, 6339-6346 (1974)

Caughey, W.S., Smythe, G.A., O'Keefee, D.H., Maskasky, J., Smith, M.L.: Heme A of cytochrome oxidase. II. Structure and properties: comparisons with hemes B, C, S and derivatives. J. Biol. Chem. 250, 7602-7622 (1975)

Cerdonio, M., Congiu-Castellano, A., Mogno, F., Pispisa, B., Romani, G.L., Vitale, S.: The magnetic properties of oxyhemoglobin. Proc. Natl. Acad. Sci. 74, 398-400 (1977)

Chance, B.: An intermediate compound in the catalase-hydrogen peroxide reaction. Acta Chem. Scand. 1, 236-267 (1947)

Chance, B.: The reaction of ferrocytochrome-c with peroxidases and peroxides. In: Enzymes and Enzyme Systems, pp. 93-104. Edsall, J.T. (ed.). Cambridge, Mass.: Harvard 1951

Chance, B., Estabrook, R., Yonetani, T.: Hemes and Hemoproteins. New York, London: Academic Press 1966

Chance, B., Saronio, C., Leigh, J.S., Jr.: Functional intermediates in reaction of cytochrome oxidase with oxygen. Proc. Natl. Acad. Sci. 72, 1635-1640 (1975)

Chance, B., Schonbaum, G.R.: Fine structure in the low temperature spectra of catalase complexes. Acta Chem. Scand. 17, S257-S262 (1963)

Chance, B., Yonetani, T., Mildvan, A.S. (eds.): Probes of enzymes and hemoproteins. Vol. II. Probes of Structure and Function of Macromolecules and Membranes. New York, London: Academic Press 1971

Cheng, J.C., Osborne, G.A., Stephens, P.J., Eaton, W.A.: Infrared magnetic circular dichroism in the study of metalloproteins. Nature (London) 241, 193-194 (1973)

Cheniae, G.M., Martin, I.F.: Effects of hydroxylamine on photosystem II. 1. Factors affecting the decay of O_2 evolution. Plant Physiol. 47, 568-575 (1971)

Cheniae, G.M., Martin, I.F.: Effects of hydroxylamine on photosystem II. 2. Photoreversal of the NH_2OH destruction of O_2 evolution. Plant Physiol. 50, 87-94 (1972)

Clementi, E.: Simple basis set for molecular wavefunctions containing first- and second-row atoms. J. Chem. Phys. 40, 1944-1945 (1964)

Clementi, E., Roothaan, C.C.J., Yoshimine, M.: Accurate analytical self-consistent field functions for atoms. II. Lowest configurations of the neutral first row atoms. Phys. Rev. 127, 1618-1620 (1962)

Coleman, J.E., Coleman, R.V.: Magnetic circular dichroism of Co (II) carbonic anhydrase. J. Biol. Chem. 247, 4718-4728 (1972)

Collman, J.P., Gagne, R.R., Reed, C.A., Halbert, T.R., Lang, G., Robinson, W.T.: "Picket fence porphyrins." Synthetic models for oxygen binding hemoproteins. J. Am. Chem. Soc. 97, 1427-1439 (1975)

Collman, J.P., Gagne, R.R., Reed, C.A., Robinson, W.T., Rodley, G.A.: Structure of an iron (II) dioxygen complex; a model for oxygen carrying hemeproteins. Proc. Natl. Acad. Sci. 71, 1326-1329 (1974)

Coryell, C.D., Stitt, F., Pauling, L.: The magnetic properties and structure of ferrihemoglobin (methemoglobin) and some of its compounds. J. Am. Chem. Soc. 59, 633-642 (1937)

Cotton, F.A., Wilkinson, G.: Advanced inorganic chemistry, 3rd ed. New York, London: Interscience 1972

Crabbé, P.: ORD and CD in Chemistry and Biochemistry, Chap. 5. New York, London: Academic Press 1972

Curzon, G.: The inhibition of caeruloplasmin by azide. Biochem. J. 100, 295-302 (1966)

Davis, L.C., Henzl, M.T., Burris, R.H., Orme-Johnson, W.H.: Iron-sulfur clusters in the molybdenum-iron protein component of nitrogenase. EPR of the CO-inhibited state. (To be published in Biochemistry)

Dawson, C.R.: Ascorbate oxidase, a review. In: The Biochemistry of Copper, pp. 305-334. Peisach, J., Aisen, P., Blumberg, W.E. (eds.). New York, London: Academic Press 1966

Dawson, J.W., Gray, H.B., Hoenig, H.E., Rossman, G.R., Schredder, J.M., Wang, R.-H.: A magnetic susceptibility study of hemerythrin using an ultrasensitive magnetometer. Biochemistry 11, 461-465 (1972)

Day, P., Smith, D.W., Williams, R.J.P.: Crystal spectra of a heme and some heme-protein complexes. Biochemistry 6, 1563-1566 (1967a)

Day, P., Smith, D.W., Williams, R.J.P.: Crystal spectra of some ferric hemoproteins. Biochemistry 6, 3747-3750 (1967b)

Dickerson, R.E., Kopka, M.L., Weinzierl, J., Varnum, J., Eisenberg, D., Margoliash, E.: Location of the heme in horse heart ferricytochrome c by X-ray diffraction. J. Biol. Chem. 242, 3015-3017 (1967)

Dickerson, R.E., Takano, T., Eisenberg, D., Kallai, O.B., Samson, L., Cooper, A., Margoliash, E.: Ferricytochrome c: I. General features of the horse and bonito proteins at 2.8 Å resolution. J. Biol. Chem. 246, 1511-1535 (1971)
Dickerson, R.E., Timkovich, R.: Cytochromes c. In: The Enzymes, Vol. XI, Chapter 7. Boyer, P.D. (ed.). New York: Academic Press 1975
Dickerson, R.E., Timkovich, R., Almassy, R.J.: The cytochrome fold and the evolution of bacterial energy metabolism. J. Mol. Biol. 100, 473-491 (1976)
Dixon, N.E., Gazzola, C., Blakeley, R.L., Zerner, B.: Metal ions in enzymes using ammonia or amides. Science 191, 1144-1150 (1976)
Drucker, H., Campbell, L.L., Woody, R.W.: Optical rotatory properties of the cytochromes c_3 from three species of *Desulfovibrio*. Biochemistry 9, 1519-1527 (1970)
Dunford, H.B., Stillman, J.S.: On the function and mechanism of action of peroxidases. Coord. Chem. Rev. 19, 187-251 (1976)
Dunham, W.R., Bearden, A.J., Salmeen, I.T., Palmer, G., Sands, R.H., Orme-Johnson, W.H., Beinert, H.: The two-iron ferredoxins in spinach, parsley, pig adrenal cortex, *Azotobacter Vinelandii*, and *Clostridium Pasteurianum:* Studies by magnetic field Mössbauer spectroscopy. Biochim. Biophys. Acta 253, 134-152 (1971a)
Dunham, W.R., Palmer, G., Sands, R.H., Bearden, A.J.: On the structure of the iron-sulfur complex in the two-iron ferredoxins. Biochim. Biophys. Acta 253, 373-384 (1971b)
Dunn, T.M., McClure, D.S., Pearson, R.G.: Some Aspects of Crystal Field Theory. New York: Harper and Row 1965
Dus, K., De Klerk, H., Sletten, K., Bartsch, R.G.: Chemical characterization of high potential iron proteins from *Chromatium* and *Rhodopseudomonas Geltinosa*. Biochim. Biophys. Acta 140, 291-311 (1967)
Dutton, P.L., Wilson, D.F., Lee, C.P.: Oxidation-reduction potentials of cytochromes in mitochondria. Biochemistry 9, 5077-5082 (1970)
Eaton, W.A., Charney, E.: Near-infrared absorption and circular dichroism spectra of ferrocytochrome c: d→d transitions. J. Chem. Phys. 51, 4502-4505 (1969)
Eaton, W.A., Hochstrasser, R.M.: Electronic spectrum of single crystals of ferricytochrome c. J. Chem. Phys. 46, 2533-2539 (1967)
Eaton, W.A., Hochstrasser, R.M.: Single-crystal spectra of ferrimyoglobin complexes in polarized light. J. Chem. Physics 49, 985-995 (1968)
Eaton, W., Palmer, G., Fee, J.A., Kimura, T., Lovenberg, W.: Tetrahedral iron in the active center of plant ferredoxins and beef adrenodoxin. Proc. Natl. Acad. Sci. 68, 3015-3020 (1971)
Ehrenberg, A.: Electron spin resonance absorption by some hemoproteins. Arkiv Kemi 19, 119-128 (1962)
Ehrenberg, A., Estabrook, R.W.: Stabilization of catalase-ammonia complex in frozen solution. Acta Chem. Scand. 20, 1667-1672 (1966)
Ehrenberg, A., Kamen, M.D.: Magnetic and optical properties of some bacterial haem proteins. Biochim. Biophys. Acta 102, 333-340 (1965)
Ehrenberg, A., Malmström, B.G., Broman, L., Mosbach, R.: A magnetic susceptibility study of copper valence in ceruloplasmin and laccase. J. Mol. Biol. 5, 450-452 (1962)
Englander, S.W.: Measurements of structural and free energy changes in hemoglobin by hydrogen exchange methods. Ann. New York Acad. Sci. 244, 10-27 (1975)
Errede, B., Haight, G.P., Jr., Kamen, M.D.: Oxidation of ferrocytochrome c by mitochondrial cytochrome c oxidase. Proc. Natl. Acad. Sci. 73, 113-117 (1976)

Fabry, T.L., Hunt, J.W., Jr.: Pressure-induced spectral shifts in hemoproteins. Arch. Biochem. Biophys. 123, 428-429 (1968)
Falk, J.E.: Porphyrins and Metalloporphyrins. Amsterdam, London, New York: Elsevier 1964
Falk, K.-E., Reinhammar, B.: Visible and near-infrared circular dichroism of some blue copper proteins. Biochim. Biophys. Acta 285, 84-90 (1972)
Falle, H.R., Luckhurst, G.R.: The temperature dependence of the electron resonance spectrum of copper acetylacetonate. Molec. Phys. 10, 597-599 (1965)
Fee, J., Malmström, B.G.: The redox potential of fungal laccase. Biochim. Biophys. Acta 153, 299-302 (1968)
Fee, J.A., Malkin, R., Malmström, B.G., Vänngård, T.: Anaerobic Oxidation-Reduction titrations of fungal laccase. Evidence for several high potential electron-accepting sites. J. Biol. Chem. 244, 4200-4207 (1969)
Feher, G., Isaacson, R.A., McElroy, J.D., Ackerson, L.C., Okamura, M.Y.: On the Question of the Primary Electron Acceptor in Bacterial Photosynthesis; Manganese Substituting for Iron in Reaction Centers of *Rhodopseudomonas sphaeroides* R-26. Biochim. Biophys. Acta 368, 135-139 (1974)
Feher, G., Malley, M., Mauzerall, D.: Direct observation of the Zeeman splitting of the excited state of porphyrins. In: Magnetic Resonance in Biological Systems, pp. 145-147. Ehrenberg, A., Malmström, B.G., Vänngård, T. (eds.). Oxford, London: Pergamon 1967
Feher, G., Richards, P.L.: Determination of the zero field splitting "D" in heme chloride by far-infrared spectroscopy. In: Magnetic Resonance in Biological Systems, pp. 141-144. Ehrenberg, A., Malmström, B.G., Vänngård, T. (eds.). Oxford, London: Pergamon 1967
Fliflet, A., Kelly, H.P.: Unpublished calculations (1973)
Florkin, M., Stotz, E.H. (eds.): Comprehensive biochemistry, Volume XIV: Biological Oxidations. Amsterdam, London, New York: Elsevier 1966
Franconi, C. (ed.): Magnetic Resonances in Biological Research. London: Gordon and Breach 1971
Freeman, H.C.: Crystal structures of metal-peptide complexes. Advan. Protein Chem. 22, 257-424 (1967)
Fridovitch, I.: Superoxide dismutases. Ann. Rev. Biochem. 44, 147-159 (1975)
Frieden, E.: The biochemical evolution of the iron and copper proteins. In: Trace Element Metabolism in Animals-2, pp. 105-118. Hoekstra, W.G., Suttie, J.W., Ganther, H.E., Mertz, W. (eds.). Baltimore, London: Butterworths 1974
Frieden, E., Osaki, S., Kobayashi, H.: Copper proteins and oxygen. J. Gen. Physiol. 49, 213-252 (1965)
Friedman, S., Kaufman, S.: An EPR study of 3,4-dihydroxyphenylethylamine β-hydroxylase. J. Biol. Chem. 241, 2256-2259 (1966)
Fritz, J., Anderson, R., Fee, J., Palmer, G., Sands, R.H., Tsibris, J.C.M., Gunsalus, I.C., Orme-Johnson, W.H., Beinert, H.: The iron electron-nuclear double resonance (ENDOR) of two-iron ferredoxins from spinach, parsley, pig adrenal cortex and *Pseudomonas Putida*. Biochim. Biophys. Acta 253, 110-133 (1971)
George, P.: On the nature of hemoprotein reactions. In: Currents in Biochemical Research, 1956, pp. 338-377. Green, D.E. (ed.). New York: Interscience 1956
George, P., Beetlestone, J., Griffith, J.S.: Ferrihaemoprotein hydroxides: A correlation between magnetic and spectroscopic properties. In: Haematin Enzymes, pp. 105-139. Falk, J.E., Lemberg, R., Morton, R.K. (eds.). New York: Pergamon 1961

George, P., Bennett, J.E., Ingram, D.J.E.: Paramagnetic resonance spectra from single crystals of hemoglobin derivatives. J. Chem. Phys. 24, 627-628 (1956)
Ghiretti, F. (ed.): Physiology and Biochemistry of Haemocyanins. New York: Academic Press 1968
Gibson, J.F., Hall, D.O., Thornley, J.H.M., Whatley, F.R.: The iron complex in spinach ferredoxin. Proc. Natl. Acad. Sci. 56, 987-990 (1966)
Gibson, J.F., Ingram, D.J.E.: Electron resonance studies of haemoglobin azide and hydroxide derivatives. Nature (London) 180, 29-30 (1957)
Goodgame, D.M.L., Hill, N.J., Marhsam, D.F., Skapski, A.C., Smart, M.L., Throughton, P.G.H.: The Insignificance of metal-metal bonding in the antiferromagnetism of copper (II) carboxylate dimers. Chem. Comm., 629-630 (1969)
Gould, D.C., Mason, H.S.: An EPR study of cupric glycylglycinate and its ethyl ester. In: The Biochemistry of Copper, pp. 35-47. Peisach, J., Aisen, P., Blumberg, W.E. (eds.). New York, London: Academic Press 1966
Gouterman, M.: Spectra of prophyrins, Part I. J. Mol. Spectr. 6, 138-163 (1961)
Gray, C.J.: Enzyme-Catalyzed Reactions. London, New York: Van Nostrand-Reinhold 1971
Griffith, J.S.: Theory of electron resonance in ferrihaemoglobin azide. Nature (London) 180, 30-31 (1957)
Griffith, J.S.: The Theory of Transition-Metal Ions. London, New York: Cambridge Univ. Press 1961
Griffith, J.S.: Theory of EPR in low-spin ferric haemoproteins. Molec. Phys. 21, 135-139 (1971)
Guzy, C.M., Raynor, J.B., Symons, M.C.R.: Electron spin resonance spectrum of copper-63 phthalocyanin. A reassessment of the bonding parameters. J. Chem. Soc. A, 2299-2303 (1969)
Hales, B.J.: Temperature dependency of the rate of electron transport as a monitor of protein motion. Biophys. J. 16, 471-480 (1976)
Hamilton, W.C., Ibers, J.A.: Hydrogen Bonding in Solids. New York, Amsterdam: Benjamin 1968
Harris, G.: Spin-mixed states of ferric ion in complexes of tetragonal symmetry. I. Eigenfunctions and eigenvalues of different spin states. Theoret. Chim. Acta 10, 119-154 (1968a)
Harris, G.: Spin-mixed states of ferric ion in complexes of tetragonal symmetry. II. Localized properties: zero field splitting, effective magnetic moments, magnetic field energies and electric field gradients. Theoret. Chim. Acta 10, 155-180 (1968b)
Harris-Loew, G.: A comparison of molecular orbital and crystal field calculations of ferric heme compounds. Theoret. Chim. Acta 17, 18-34 (1970)
Hartree, D.R.: Calculation of Atomic Structures. New York: Wiley 1957
Hartree, E.F.: Magnetic properties of haematin derivatives. Ann. Rep. Progr. Chem. 43, 287-296 (1946)
Hartsuck, J.A., Lipscomb, W.N.: Carboxypeptidase A. In: The Enzymes, Vol. 3, Chap. 1. Boyer, P. (ed.). New York: Academic Press 1971
Haurowitz, F., Clar, E., Hermann, Z., Kittel, H., Münzberg, F.K.: Eigenschaften der Porphyrin-Metall-Komplexe und ihre Abhängigkeit von der Wertigkeit und Susceptibilität ihrer Metallatome. Ber. Deut. Chem. Ges. 68, 1795-1806 (1935)
Hayaishi, A., Shimizu, A., Yamamura, Y., Watari, H.: Hemoglobins M: identification of Iwate, Boston, and Saskatoon variants. Science 152, 207-208 (1966)
Hayaishi, O.: Enzymatic Hydroxylation. Ann. Rev. Biochem. 38, 21-44 (1969)

Helcké, G.A., Ingram, D.J.E., Slade, E.F.: Electron resonance studies of hemoglobin derivatives. III. Line-width and g-value measurements of acid-metmyoglobin and of metmyoglobin azide derivatives. Proc. Roy. Soc. (London), Ser. B 169, 275-288 (1968)
Hendrickson, W.A., Klippenstein, G.L., Ward, K.B.: Tertiary structure of myohemerythrin at low resolution. Proc. Natl. Acad. Sci. 72, 2160-2164 (1975)
Herskovitz, T., Averill, B.A., Holm, R.H., Ibers, J.A., Phillips, W.D., Weiher, J.F.: Structure and properties of a synthetic analogue of bacterial iron-sulfur proteins. Proc. Natl. Acad. Sci. 69, 2437-2441 (1972)
Hill, N.J.: Unpublished experiments (1972)
Hoard, J.L.: Some aspects of heme stereochemistry. In: Structural Chemistry and Molecular Biology, pp. 573-594. Rich, A., Davidson, N. (eds.). San Francisco, London: Freeman 1968
Hoard, J.L.: Stereochemistry of hemes and other metalloporphyrins. Science 174, 1295-1302 (1971)
Hochstrasser, R.M.: Structural sensitive aspects of the electronic spectrum. In: Probes of Structure and Function of Macromolecules and Membranes. Vol. I, pp. 57-64. Chance, B.C., Lee, C.-P., Blasie, J.K. (eds.). New York: Academic Press 1971
Hodges, H.L., Holwerda, R.A., Gray, H.B.: Kinetic studies of the reduction of ferricytochrome c by $Fe(EDTA)^{2-}$. J. Am. Chem. Soc. 96, 3132-3137 (1974)
Hoffman, D.M., Petering, D.H.: Coboglobins: oxygen-carrying cobalt-reconstituted hemoglobin and myoglobin. Proc. Natl. Acad. Sci. 67, 637-643 (1970)
Hopfield, J.J.: Electron transfer between biological molecules by thermally activated tunneling. Proc. Natl. Acad. Sci. 71, 3640-3644 (1974)
Hopfield, J.J.: On electron transfer. Biophys. J. 16, 1239-1240 (1976)
Hori, H.: Analysis of the principal g-tensors in single crystals of ferrimyoglobin complexes. Biochim. Biophys. Acta 251, 227-235 (1971)
Horrocks, W. DeW., Jr., Greenberg, E.S.: Evaluation of dipolar nuclear magnetic resonance shifts in low-spin hemin systems: Ferricytochrome c and metmyoglobin Cyanide. Biochim. Biophys. Acta 322, 38-44 (1973)
Hsu, M.-C., Woody, R.W.: The origin of the heme cotton effects in myoglobin and hemoglobin. J. Am. Chem. Soc. 93, 3515-3525 (1971)
Hsu, Y.: E.P.R. studies of three cupric complexes with nitrogenous ligands. J. Molec. Phys. 21, 1087-1103 (1971)
Huber, R., Epp, O., Formanek, H.: Structures of deoxy- and carbonmonoxy-erythrocruorin. J. Mol. Biol. 52, 349-354 (1970)
Hughes, M.N.: The Inorganic Chemistry of Biological Processes. London, New York, Sydney, Toronto: Wiley-Interscience 1975
Huynh, B.H., Papaefthymiou, G.C., Yen, C.S., Groves, J.L., Wu, C.S.: Electronic structure of Fe^{2+} in normal human hemoglobin and its isolated units. J. Chem. Phys. 61, 3750-3758 (1974)
Iizuka, T., Kotani, M.: Analysis of a thermal equilibrium phenomenon between high-spin and low-spin states of ferrimyoglobin azide. Biochim. Biophys. Acta 154, 417-419 (1968)
Iizuka, T., Kotani, M.: Analysis of thermal equilibrium between high-spin and low-spin states in ferrihemoglobin complexes. Biochim. Biophys. Acta 194, 351-363 (1969)
Iizuka, T., Yonetani, T.: Spin changes in hemoproteins. Advan. Biophys. 1, 157-182 (1970)
Ilan, Y., Shafferman, A., Stein, G.: The study of 1-electron equivalent oxidation-reduction reactions by fast pulse generation of reagents. J. Biol. Chem. 251, 4336-4345 (1976)
Ingram, D.J.E.: Biological and Biochemical Applications of Electron Spin Resonance. New York: Plenum Press 1969

Jirgensons, B.: Optical Activity of Proteins and Other Molecules. Berlin-Heidelberg-New York: Springer 1973

Jortner, J.: Temperature dependent activation energy for electron transfer between biological molecules. J. Chem. Phys. 64, 4860-4867 (1976)

Kasper, C.B.: Ceruloplasmin-anion interactions. Induced spectral transitions in the visible range. J. Biol. Chem. 243, 3218-3222 (1968)

Keele, B.B., Jr., McCord, J.M., Fridovitch, I.: Superoxide dismutase from *Escherichia coli* B. A new manganese-containing enzyme. J. Biol. Chem. 245, 6176-6181 (1970)

Keele, B.B., Jr., McCord, J.M., Fridovitch, I.: Further characterization of bovine superoxide dismutase and its isolation from bovine heart. J. Biol. Chem. 246, 2875-2880 (1971)

Keilin, D., Hartree, E.F.: The combination between methaemoglobin and peroxides: hydrogen peroxide and ethyl hydroperoxide. Proc. Roy. Soc. (London), Ser. B 117, 1-15 (1935)

Kendrew, J.C.: Myoglobin and the structure of proteins. Science 139, 1259-1266 (1963)

Kendrew, J.C., Dickerson, R.E., Strandberg, B.E., Hart, R.G., Davies, D.R., Phillips, D.C., Shore, V.C.: Structure of myoglobin. Nature (London) 185, 422-427 (1960)

Kendrew, J.C., Parrish, R.G.: The crystal structure of myoglobin. III. Sperm-whale myoglobin. Proc. Roy. Soc. (London), Ser. A 238, 305-324 (1956)

Kincaid, B.B., Eisenberger, P., Hodgson, K.O., Doniach, S.: X-ray absorption spectroscopy using synchrotron radiation for structural investigation of organometallic molecules of biological interest. Proc. Natl. Acad. Sci. 72, 2340-2342 (1975)

King, T.E., Mason, H.S., Morrison, M. (eds.): Oxidases and Related Redox Systems, 2 vols. Baltimore: University Park Press 1973

Kivelson, D., Nieman, R.: ESR studies on the bonding in copper complexes. J. Chem. Phys. 35, 149-155 (1961)61)

Kobayashi, H.: On the electronic spectra of porphine. J. Chem. Phys. 30, 1362-1363 (1959)

Kokoszka, G.F., Gordon, G.: Metal-metal exchange interactions. In: Transition Metal Chemistry, Vol. V, pp. 181-277. Carlin, R.L. (ed.). New York: Dekker 1969

Kotani, M.: Paramagnetic properties and electronic structure of iron in heme proteins. Advan. Quantum Chem. 4, 227-266 (1968)

Kotani, M., Morimoto, H.: EPR studies on single crystals of myoglobin and myoglobin fluoride. In: Magnetic Resonance in Biological Systems, pp. 135-140. Ehrenberg, A., Malmström, B.G., Vänngård, T. (eds.). Oxford, London: Pergamon 1967

Kretsinger, R.H.: Hypothesis: calcium modulated proteins contain EF hands. In: Calcium Transport in Contraction and Secretion, pp. 469-478. Carafoli, E. (ed.). Amsterdam: North-Holland 1975

Kuska, H.A., Rogers, M.T.: Electron spin resonance of coordination compounds. In: Spectroscopy in Inorganic Chemistry, Vol. II, pp. 175-196, Rao, C.N.R., Ferraro, J.R. (eds.). New York: Academic Press 1971

Lang, G.: Mössbauer spectroscopy of haem proteins. Quart. Rev. Biophys. 3, 1-60 (1970)

Lang, G., Marshall, W.: Mössbauer effect in some haemoglobin compounds. Proc. Phys. Soc. 87, 3-34 (1966)

Lehninger, A.L.: Biochemistry. 2nd ed. New York: Worth 1975

Lemberg, R., Barrett, J.: Cytochromes. New York: Academic Press 1973

Lemberg, R., Legge, J.W.: Hematin compounds and bile pigments. New York: Interscience 1949

Loew, G.H.: An analysis of the electron spin resonance of low spin ferric heme compounds. Biophys. J. 10, 196-212 (1970)
Loew, G.H.: Nonlinear magnetic behavior of five high-spin ferric heme compounds. J. Mag. Res. 6, 408-421 (1972)
Lovenberg, W. (ed.): Iron-Sulfur Proteins. Vol. I. Biological Properties, Vol. II, Molecular Properties. New York: Academic Press 1973
Maehly, A.C.: Haemin-protein binding in peroxidase and methaemalbumin. Nature (London) 192, 630-632 (1961)
Mailer, C., Taylor, C.P.S.: Electron paramagnetic resonance study of single crystals of horse heart ferricytochrome c at 4.2 °K. Can. J. Biochem. 50, 1048-1055 (1972)
Maki, A.H., McGarvey, B.R.: Electron spin resonance in transition metal chelates. II. Copper (II) Bis-salicylaldehyde-imine. J. Chem. Phys. 29, 35-38 (1958)
Malkin, R., Malmström, B.G.: The state and function of copper in biological systems. Advan. Enzymol. 33, 177-244 (1970)
Malkin, R., Malmström, B.G., Vänngård, T.: The requirement of the "non-blue" copper (II) for the activity of fungal laccase. FEBS Lett. 1, 50-54 (1968)
Malkin, R., Malmström, B.G., Vänngård, T.: The reversible removal of one specific copper (II) from fungal Laccase. Europ. J. Biochem. 7, 253-259 (1969a)
Malkin, R., Malmström, B.G., Vänngård, T.: Spectroscopic differentiation of the electron-accepting sites in fungal laccase. Europ. J. Biochem. 10, 324-329 (1969b)
Malmström, B.G., Finazzi-Agrò, A., Antonini, E.: The mechanisms of laccase-catalyzed oxidations: kinetic evidence for the involvement of several electron-accepting sites in the enzyme. Europ. J. Biochem. 9, 383-391 (1969)
Malmström, B.G., Reinhammar, B., Vänngård, T.: Two forms of copper (II) in fungal laccase. Biochim. Biophys. Acta 156, 67-76 (1968)
Malmström, B.G., Reinhammar, B., Vänngård, T.: The state of copper in stellacyanin and laccase from the lacquer tree *Rhus Vernicifera*. Biochim. Biophys. Acta 205, 48-57 (1970)
Malmström, B.G., Vänngård, T.: Electron spin resonance of copper proteins and some model complexes. J. Mol. Biol. 2, 118-124 (1960)
Maltempo, M.M.: Magnetic state of an unusual bacterial heme protein. J. Chem. Phys. 61, 2540-2547 (1974)
Maltempo, M.M., Moss, T.H.: The spin 3/2 state and quantum spin mixtures in haem proteins. Quart. Rev. Biophys. 9, 181-215 (1976)
Maltempo, M.M., Moss, T.H., Cusanovich, M.A.: Magnetic studies on the changes in the iron environment in *Chromatium* ferricytochrome c'. Biochim. Biophys. Acta 342, 290-305 (1974)
Margoliash, E., Schejter, A.: Kinetics of the irreversible inhibition of catalase by 3-amino-1,2,4-triazole in the presence of hydrogen peroxide and catalase-hydrogen peroxide complex I hydrogen donors. J. Biol. Chem. 237, 2359-2363 (1962)
Margoliash, E., Schejter, A.: Cytochrome c. Advan. Protein Chem. 21, 113-286 (1966)
Maria, H.J.: Colour, magnetic resonance, and symmetry of the cupric site in blue protein. Nature (London) 209, 1023-1024 (1966)
Markley, J.L., Ulrich, E.L., Berg, S.P., Krogmann, D.W.: Nuclear magnetic resonance studies of the copper binding sites of blue copper proteins: oxidized, reduced, and apoplastocyanin. Biochemistry 14, 4428-4433 (1975)
Martin, R.B.: Personal communication, 1973

Martin, R.B.: Optical properties of transition metal ion complexes of amino acids and peptides. In: Metal Ions in Biological Systems. Vol. I, Chap. 4. Sigel, H. (ed.). New York: Dekker 1974

Mason, H.S.: An ESR study of *Pseudomonas* copper protein. Biochem. Biophys. Res. Commun. 10, 11-13 (1963)

Mathews, F.S., Levine, M., Argos, P.: The structure of calf liver cytochrome b_5 at 2.5 Å resolution. Nature (London) 233, 15-16 (1971a)

Mathews, F.S., Levine, M., Argos, P.: X-ray study of calf liver cytochrome b_5. In: Probes of Structure and Function of Macromolecules and Membranes, Vol. II, pp. 505-510. Chance, B., Yonetani, T., Mildvan, A. (eds.). New York: Academic Press 1971b

Mathews, R., Charlton, S., Sands, R.H., Palmer, G.: On the nature of the spin coupling between the iron-sulfur clusters in the eight-iron ferredoxins. J. Biol. Chem. 13, 4326-4328 (1974)

McCord, J.M., Fridovitch, I.: Superoxide dismutase. An enzymic function for erythrocuprein (hemocuprein). J. Biol. Chem. 244, 6049-6055 (1969)

McMillan, D.R., Holwerda, R.A., Gray, H.B.: Preparation and spectroscopic studies of cobalt (II)-stellacyanin. Proc. Natl. Acad. Sci. 71, 1339-1341 (1974)

Meriwether, L.S., Marzluff, W.F., Hodgson, W.G.: Molybdenum-thiol complexes as models for molybdenum bound in enzymes. Nature (London) 212, 465-467 (1966)

Michel-Calendini, F.M., Kibler, M.R.: Fonctions autocohérentes et intégrales radiales pour les ions de la série du fer. Theoret. Chim. Acta. 10, 367-371 (1968)

Mildvan, A.S., Cohn, M.: Aspects of enzyme mechanisms studied by nuclear spin relaxation induced by paramagnetic probes. Advan. Enzymol. 33, 1-70 (1970)

Miskowski, V., Tang, S.-P.W., Spiro, T.G., Shapiro, E., Moss, T.H.: The copper coordination group in "blue" copper proteins: Evidence from resonance Raman spectra. Biochemistry 14, 1244-1250 (1975)

Moore, G.R., Williams, R.J.P.: Electron-transfer proteins. Coord. Chem. Rev. 18, 125-197 (1976)

Morita, Y., Mason, H.S.: An electron spin resonance study of some hemoproteins. J. Biol. Chem. 240, 2654-2659 (1965)

Morpurgo, L., Finazzi-Agrò, A., Rotilio, G., Mondavì, B.: IV. Stellacyanin: Preparation of apoprotein and involvement of sulfhydryl and tryptophan in the copper chromophore. Biochim. Biophys. Acta 271, 292-299 (1972)

Moss, T.H.: Mössbauer spectroscopy. Meth. Enzymol. 27, 912-941 (1973)

Moss, T.H., Moleski, C., York, J.L.: Magnetic susceptibility evidence for a binuclear iron complex in hemerythrin. Biochemistry 10, 840-842 (1971)

Muirhead, H., Greer, J.: Three-dimensional Fourier Synthesis of human deoxyhaemoglobin at 3.5 Å resolution. Nature (London) 228, 516-519 (1970)

Murphy, M.J., Siegel, L.M., Kamin, H., Rosenthal, D.: Reduced nicotinamide adenine dinucleotide phosphate-sulfite reductase of enterobacteria. II. Identification of a new class of heme prosthetic group. J. Biol. Chem., 2801-2814 (1973)

Norvell, J.C., Nunes, A.C., Schoenborn, B.P.: Neutron diffraction analysis of myoglobin: structure of the carbon monoxide derivative. Science 190, 568-570 (1975)

O'Dell, B.L., Campbell, B.J.: Trace Elements: metabolism and metabolic function. In: Comprehensive Biochemistry, Vol. XXI, Chap. 2. Florkin, M., Stotz, E.H. (eds.). Amsterdam, London, New York: Elsevier 1970

Offenhartz, P.O'D.: Electronic spectra of ferrimyoglobin hydroxide and other hemoprotein derivatives. J. Chem. Phys. 42, 3566-3572 (1965)

Ogunmola, G.B., Kauzmann, W., Zipp, A.: Volume changes in binding of ligands to methemoglobin and metmyoglobin. Proc. Natl. Acad. Sci. 73, 4271-4273 (1976)

Okamura, M.Y., Klotz, I.M., Johnson, C.E., Winter, M.R.C., Williams, R.J.P.: The state of hemerythrin. A Mössbauer study. Biochemistry 8, 1951-1958 (1969)
Okunuki, K.: Cytochromes and cytochrome oxidase. In: Comprehensive Biochemistry, Vol. XIV, Chap. 5. Florkin, M., Stotz, E.H. (eds.). Amsterdam, London, New York: Elsevier 1966
Olson, J.M., Chance, B.C.: Oxidation-reduction reactions in the photosynthetic bacterium *Chromatium*. II. Dependence of light reactions on intensity of irradiation and quantum efficiency of cytochrome oxidation. Arch. Biochem. Biophys. 88, 40-53 (1960)
Oosterhuis, W.T.: The electronic state of iron in some natural iron compounds: Determination by Mössbauer and ESR spectroscopy. Struct. Bonding 20, 59-99 (1974)
Orme-Johnson, W., Beinert, H.: Reductive titrations of iron-sulfur proteins containing two to four iron atoms. J. Biol. Chem. 244, 6143-6148 (1969)
Osaki, S.: Discussion of bonding in copper chromophores. In: The Biochemistry of Copper, pp. 13-14. Peisach, J., Aisen, P., Blumberg, W.E. (eds.). New York, London: Academic Press 1966
Owen, J., Thornley, J.H.M.: Covalent bonding and magnetic properties of transition metal ions. Repts. Prog. Phys. 29, 675-728 (1966)
Padlan, E.A., Love, W.E.: Alterations in the projected electron density of *Glycera* hemoglobin accompanying changes in ligand state. In: Probes of Structure and Function of Macromolecules and Membranes. Vol. II, pp. 187-192. Chance, B.C., Yonetani, T., Mildvan, A.S. (eds.). New York: Academic Press 1971
Palmer, G.: Iron-sulfur proteins. In: The Enzymes, Vol. XII, Chap. 1. Boyer, P.D. (ed.). New York, Academic Press 1975
Palmer, G., Dunham, W.R., Fee, J.A., Sands, R.H., Iizuka, T., Yonetani, T.: The magnetic susceptibility of spinach ferredoxin from 77-250 ^{o}K: A measurement of the antiferromagnetic coupling between the two iron atoms. Biochim. Biophys. Acta 245, 201-207 (1971)
Paul, K.G.: Artificial peroxidases. Acta Chem. Scand. 13, 1239-1242 (1959)
Paul, K.G., Stigbrand, T.: Umecyanin, a novel intensely blue copper protein from horseradish root. Biochim. Biophys. Acta 221, 255-263 (1970)
Peisach, J., Aisen, P., Blumberg, W.E. (eds.). The Biochemistry of Copper. New York, London: Academic Press 1966
Peisach, J., Blumberg, W.E., Ogawa, S., Rachmilewitz, E.A., Oltzik, R.: The effects of protein conformation on the heme symmetry in high spin ferric heme proteins as studied by electron paramagnetic resonance spectroscopy. J. Biol. Chem. 246, 3342-3355 (1971)
Peisach, J., Levine, W.G., Blumberg, W.E.: Structural properties of Stellacyanin, a copper mucoprotein from *Rhus vernicifera*, the Japanese Lac tree. J. Biol. Chem. 242, 2847-2858 (1967)
Perutz, M.F.: Polarization dichroism, form birefringence, and molecular orientation in crystalline hemoglobins. Acta Cryst. 6, 859-864 (1953)
Perutz, M.F.: Stereochemistry of cooperative effects in haemoglobin. Nature (London) 228, 726-739 (1970)
Phillips, C.S.G., Williams, R.J.P.: Inorganic Chemistry, Vol. II. New York, Oxford: Oxford Univ. Press 1966
Phillips, W.D.: The nuclear magnetic resonance spectroscopy of proteins. In: Probes of Structure and Function of Macromolecules and Membranes, Vol. I, pp. 75-95. Chance, B.C., Lee, C.-P., Blasie, J.K. (eds.). New York: Academic Press 1971
Phillips, W.D.: Biological applications. In: NMR of Paramagnetic Molecules: Principles and Applications, pp. 421-478. La Mar, G.N., Horrocks, W.DeW., Jr., Holm, R.H. (eds.). New York: Academic Press 1973

Platt, J.: Electronic structure and excitation of polyenes and porphyrins. In: Radiation Biology, Vol. III, Chap. 2, pp. 71-123. Hollaender, A. (ed.). New York: McGraw-Hill 1956

Poe, M., Phillips, W.D., Glickson, J.D., San Pietro, A.: Proton magnetic resonance studies of the ferredoxins from spinach and parsley. Proc. Natl. Acad. Sci. 68, 68-71 (1971)

Poe, M., Phillips, W.D., McDonald, C.C., Lovenberg, W.: Proton magnetic resonance study of ferredoxin from *Clostridium pasteurianum*. Proc. Natl. Acad. Sci. 65, 797-804 (1970)

Redfield, A.G., Gupta, R.K.: Pulsed NMR study of the structure of cytochrome c. Cold Spr. Harbor Symp. Quant. Biol. 36, 405-411 (1972)

Rein, H., Ruckpaul, K., Haberditzl, W.: Magneto-optical rotation spectra of methemoglobin in the presence of allosteric effectors. FEBS Lett. 32, 166-170 (1973)

Ricard, J., Mazza, G., and (in part) Williams, R.J.P.: Oxidation-reduction potentials and ionization states of two turnip peroxidases. European J. Biochem. 28, 566-578 (1972)

Richards, P.L., Caughey, W.S., Eberspaecher, H., Feher, G., Malley, M.: Determination of the zero-field splitting of Fe^{3+} in several hemin compounds. J. Chem. Phys. 47, 1187-1188 (1967)

Richardson, F.S.: Optical activity of transition metal compounds. I. Sector rules for metal complexes of the pseudotetragonal class. J. Chem. Phys. 54, 2453-2468 (1971)

Richardson, J.S., Thomas, K.A., Rubin, B.H., Richardson, D.C.: Crystal structure of bovine Cu, Zn superoxide dismutase at 3 Å resolution: chain tracing and metal ligands. Proc. Natl. Acad. Sci. 72, 1349-1353 (1975)

Richardson, J.W., Nieuwport, W.C., Powell, R.R., Edgell, W.F.: Approximate radial functions for first-row transition-metal atoms and ions. I. Inner-shell, 3d and 4s atomic orbitals. J. Chem. Phys. 36, 1057-1061 (1962)

Richardson, J.W., Powell, R.R., Nieuwport, W.C.: Approximate radial functions for first-row transition-metal atoms and ions. II. 4p and 4d atomic orbitals. J. Chem. Phys. 38, 796-801 (1963)

Risler, J.-L., Groudinsky, O.: Magnetic-circular dichroism studies of cytochrome c and cytochrome b_2. Europ. J. Biochem. 35, 201-205 (1973)

Rist, G.H., Hyde, J.S., Vänngård, T.: Electron-nuclear double resonance of a protein that contains copper: evidence for nitrogen coordination to Cu (II) in stellacyanin. Proc. Natl. Acad. Sci. 67, 79-86 (1970)

Roberts, E.M., Koski, W.S.: Electron spin resonance of copper etioporphyrin II. J. Am. Chem. Soc. 82, 3006-3010 (1960)

Rotilio, G., Bray, R.C., Fielden, E.M.: A pulse radiolysis study of superoxide dismutase. Biochim. Biophys. Acta 268, 605-609 (1972)

Roughton, F.J.W., Chance, B.: Rapid reactions. In: Techniques of Organic Chemistry, Vol. VIII, part 2, pp. 703-792. Weissberger, A. (ed.). New York, London: Interscience, 1963

Ruckpaul, K., Rein, H., Jung, F.: Zirkulardichroismus von Porphyrin-Globinen. Naturwissenschaften 57, 131-132 (1970)

Salemme, F.R., Freer, S.T., Xuong, NG.H., Alden, R.A., Kraut, J.: The structure of oxidized cytochrome c_2 of *Rhodospirillum rubrum*. J. Biol. Chem. 248, 3910-3921 (1973)

Sandberg, H.E., Balegh, M.S.: Evidence for the coordination of a histidyl residue to heme. I. Far ultraviolet spectral studies of model complexes. Biochim. Biophys. Acta 295, 37-48 (1973)

Sands, R.H., Dunham, W.R.: Spectroscopic studies on two-iron ferredoxins. Quart. Rev. Biophys. 7, 443-504 (1974)

Schatz, P.N., McCaffery, A.J.: The Faraday Effect. Quart. Rev. (London) 23, 552-584 (1969)

Scheler, W., Schoffa, G., Jung, F.: Lichtabsorption und paramagnetische Suszeptibilität bei Derivaten des Pferde- und Chironomous-Methämoglobins sowie des Pferde-Metmyoglobins. Biochem. Z. 329, 232-246 (1957)
Schellman, J.A.: Symmetry rules for optical rotation. J. Chem. Phys. 44, 55-63 (1966)
Schoenborn, B.P.: A neutron diffraction analysis of myoglobin. III. Hydrogen-deuterium bonding in side chains. Cold Spr. Harb. Symp. Quant. Biol. 36, 569-575 (1971)
Scholes, C.P.: Heme oriented in an organic crystalline environment, Appendix II, Ph.D. dissertation. Yale Univ. 1969
Scholes, C.P.: EPR studies on heme oriented in an organic crystalline environment. J. Chem. Phys. 52, 4890-4895 (1970)
Scholes, C.P., Isaacson, R.A., Feher, G.: Determination of the zero-field splitting of Fe^{3+} in heme proteins. Biochim. Biophys. Acta 244, 206-210 (1971)
Scholes, C.P., Isaacson, R.A., Feher, G.: Electron nuclear double resonance studies on heme proteins: determination of the interaction of Fe^{3+} with its ligand nitrogens in metmyoglobin. Biochim. Biophys. Acta 263, 448-452 (1972)
Scholes, C.P., Van Camp, H.L.: Endor from nitrogens and protons in low spin ferric heme and hemoprotein. Biochim. Biophys. Acta 434, 290-296 (1976)
Schoot Uiterkamp, A.J.M., Mason, H.S.: Magnetic dipole-dipole coupled Cu (II) pairs in NO-treated tyrosinase: A structural relationship between the active sites of tyrosinase and hemocyanin. Proc. Natl. Acad. Sci. 70, 993-996 (1973)
Schugar, H.J., Ou, C., Thich, J.A., Potenza, J.A., Lalancette, R.A., Furey, W., Jr.: Molecular structure and copper (II)-mercaptide charge-transfer spectra of a novel $Cu_{14}[SC(CH_3)_2\ CH_2\ NH_2]_{12}\ Cl$ cluster. J. Am. Chem. Soc. 98, 3047-3048 (1976)
Setlow, R.B., Pollard, E.C.: Molecular Biophysics. Reading, Palo Alto, London: Addison-Wesley 1962
Shafferman, A., Stein, G.: Reduction of ferricytochrome c by some free radical agents. Science 183, 428-429 (1974)
Shashoua, V.E.: Magneto-optical rotation spectra of cytochrome c. Nature (London) 203, 972-973 (1964)
Shashoua, V.E.: Magneto-optical rotation spectra of porphyrins and phthalocyanines. J. Am. Chem. Soc. 87, 4044-4048 (1965)
Shethna, Y.I., Wilson, P.W., Hansen, R.E., Beinert, H.: Identification by isotopic substitution of the EPR signal at g = 1.94 in a non-heme iron protein from *Azotobacter*. Proc. Natl. Acad. Sci. 52, 1263-1271 (1964)
Shulman, R.G., Glarum, S.H., Karplus, M.: Electronic structure of cyanide complexes of hemes and heme proteins. J. Mol. Biol. 57, 93-115 (1971)
Shulman, R.G., Sugano, S.: Molecular orbital analysis of iron-group cyanides. J. Chem. Phys. 42, 39-43 (1965)
Shulman, R.G., Yafet, Y., Eisenberger, P., Blumberg, W.E.: Observation and interpretation of X-ray absorption edges in iron compounds and proteins. Proc. Natl. Acad. Sci. 73, 1384-1388 (1976)
Sieker, L.C., Adman, E., Jensen, L.H.: Structure of the Fe-S complex in a bacterial ferredoxin. Nature (London) 235, 40-42 (1972)
Simpson, W.T.: Electronic states of organic molecules. J. Chem. Phys. 16, 1124-1136 (1948)
Slater, J.C.: Atomic shielding constants. Phys. Rev. 36, 57-64 (1930)
Smith, D.W., Williams, R.J.P.: The spectra of ferric haems and haemoproteins. Struct. Bonding 7, 1-45 (1970)
Smith, L., Conrad, H.E.: A study of the kinetics of the oxidation of cytochrome c by cytochrome c oxidase. Arch. Biochem. Biophys. 63, 403-413 (1956)

Smith, L., Davies, H.C., Reichlin, M., Margoliash, E.: Separate oxidase and reductase reaction sites on cytochrome c demonstrated with purified site-specific antibodies. J. Biol. Chem. 248, 237-243 (1973)
Solomon, E.I., Clendening, P.J., Gray, H.B., Grunthaner, F.J.: Direct observation of sulfur coordination in bean plastocyanin by X-ray photoelectron spectroscopy. J. Am. Chem. Soc. 97, 3878-3879 (1975)
Sono, M., Asakura, T.: Decrease in oxygen affinity of myoglobin by formylation of vinyl groups of heme. J. Biol. Chem. 250, 5227-5232 (1975)
Speyer, B.E., Curzon, G.: The inhibition of caeruloplasmin by cyanide. Biochem. J. 106, 905-911 (1968)
Spiro, T.G.: Biological applications of resonance Raman spectroscopy: Haem proteins. Proc. Roy. Soc. (London), Ser. A 345, 89-105 (1975)
Spiro, T.G., Strekas, T.C.: Resonance raman spectra of hemoglobin and cytochrome c: inverse polarization and vibronic scattering. Proc. Natl. Acad. Sci. 69, 2622-2626 (1972)
Springall, J., Stillman, M.J., Thomson, A.J.: Low temperature magnetic circular dichroism spectra of met- and myoglobin derivatives. Biochim. Biophys. Acta 453, 494-501 (1976)
Steitz, T.A., Ludwig, M.L., Quiocho, F.A., Lipscomb, W.N.: The structure of carboxypeptidase A. V. Studies of enzyme-substrate and enzyme-inhibitor complexes at 6 Å resolution. J. Biol. Chem. 242, 4662-4668 (1967)
Stellwagen, E.: Location of the heme moiety of cytochrome c by solvent perturbation. J. Biol. Chem. 242, 602-606 (1967)
Stenkamp, R.E., Sieker, L.C., Jensen, L.H.: Structure of the iron complex in methemerythrin. Proc. Natl. Acad. Sci. 73, 349-351 (1976a)
Stenkamp, R.E., Sieker, L.C., Jensen, L.H., Loehr, J.S.: Structure of methemerythrin at 5 Å resolution. J. Mol. Biol. 100, 23-34 (1976b)
Stephens, P.J., Svëtaak, W., Shatz, P.N.: Magneto-optical rotatory dispersion of porphyrins and phthalocyanines. J. Chem. Phys. 44, 4592-4602 (1966)
Stern, E.A.: Theory of the extended X-ray absorption fine structure. Phys. Rev. B10, 3027-3037 (1974)
Stevens, K.W.H.: On the magnetic properties of covalent XY_6 complexes. Proc. Roy. Soc. A 219, 542-555 (1953)
Stigbrand, T.: Structural properties of umecyanin. Biochim. Biophys. Acta 236, 246-252 (1971)
Stigbrand, T., Malmström, B.G., Vänngård, T.: On the state of copper in the blue protein umecyanin. FEBS Lett. 12, 260-262 (1971)
Stigbrand, T., Sjöholm, I.: Circular dichroism studies on the copper protein umecyanin. Biochim. Biophys. Acta 263, 244-257 (1972)
Stryer, L., Kendrew, J.C., Watson, H.C.: The mode of attachment of the azide ion to sperm whale metmyoglobin. J. Mol. Biol. 8, 96-104 (1964)
Sugita, Y., Nagai, M., Yoneyama, Y.: Circular dichroism of hemoglobin in relation to the structure surrounding the heme. J. Biol. Chem. 246, 383-388 (1971)
Sundberg, R.J., Martin, R.B.: Interactions of histidine and other imidazole derivatives with transition metal ions in chemical and biological systems. Chem. Rev. 74, 471-517 (1974)
Sutherland, J.C., Axelrod, D., Klein, M.P.: Zeeman effect in porphyrins: zero-field splitting of the excited electronic states. J. Chem. Phys. 54, 2888-2898 (1971)
Sutherland, J.C., Klein, M.P.: Magnetic circular dichroism of cytochrome c. J. Chem. Phys. 57, 76-86 (1972)
Swartz, H.M., Bolton, J.R., Borg, D.C. (eds.): Biological Applications of Electron Spin Resonance. New York, London: Wiley-Interscience 1972
Tamura, M.: Optical and magnetic measurements of horseradish peroxidase. I. Azide complex of peroxidase. Biochim. Biophys. Acta 243, 239-248 (1971a)

Tamura, M.: Optical and magnetic measurements of horseradish peroxidase. II. pH dependence of peroxidase. Biochim. Biophys. Acta 243, 249-258 (1971b)
Tanford, C.: Physical Chemistry of Macromolecules. New York, London: Wiley 1961
Tang, S.-P.W., Coleman, J.E., Meyer, Y.P.: Conformational studies of copper-proteins. J. Biol. Chem. 243, 4286-4297 (1968)
Tang, S.-P.W., Spiro, T.G., Antanaitis, C., Moss, T.H., Holm, R.H., Herskovitz, T., Mortensen, L.E.: Resonance Raman spectroscopic evidence for structural variation among bacterial ferredoxin, HiPiP, and $Fe_4S_4(SCH_2Ph)_4{}^{2-}$. Biochem. Biophys. Res. Commun. 62, 1-6 (1975)
Tasaki, A., Ōtsuka, J., Kotani, M.: Magnetic susceptibility measurements on hemoproteins down to 4.2 °K. Biochim. Biophys. Acta 140, 284-290 (1967)
Taube, H.: Electron transfer reactions of complex ions in solution. New York: Academic Press 1970
Theorell, H.: The magnetic properties of crystalline horseradish peroxidase and some of its derivatives. Arkiv Kemi, Min. Geol. 16, A, No. 3, 1-11 (1942)
Theorell, H., Ehrenberg, A.: Spectrophotometric, magnetic and titrimetric studies on the heme-linked groups in myoglobin. Acta Chem. Scand. 5, 823-848 (1951)
Thomson, A.J., Brittain, T., Greenwood, C., Springall, J.: Determination of the heme spin states in cytochrome c oxidase using magnetic circular dichroism. FEBS Lett. 67, 94-98 (1976)
Tinkham, M.: Group Theory and Quantum Mechanics. New York: McGraw-Hill 1964
Tolansky, S., Forester, G.O.: The nuclear magnetic moment of copper. Proc. Phys. Soc. A 50, 826-833 (1938)
Treu, J.I., Hopfield, J.J.: Magnetic circular dichroism in hemoglobin. J. Chem. Phys. 63, 613-623 (1975)
Tsangaris, J.M., Martin, R.B.: Visible circular dichroism of copper (II) complexes of amino acids and peptides. J. Am. Chem. Soc. 92, 4255-4260 (1970)
Tsibris, J.C.M., Woody, R.W.: Structural studies of iron-sulfur proteins. Coord. Chem. Rev. 5, 417-458 (1970)
Uenoyama, H., Iizuka, T., Morimoto, H., Kotani, M.: Paramagnetic anisotropy measurements on acid ferrimyoglobin and ferrimyoglobin fluoride. Biochim. Biophys. Acta 160, 159-166 (1968)
Ulmer, D.D.: Optical rotatory dispersion of oxidized and reduced cytochrome c. Biochemistry 4, 902-907 (1965)
Urry, D.W., Doty, P.: On the conformation of horse heart ferri- and ferrocytochrome c. J. Am. Chem. Soc. 87, 2756-2758 (1965)
Vallee, B.L., Coleman, J.E.: Metal coordination and enzyme action. In: Comprehensive Biochemistry, Vol. XII, Chap. 6. Florkin, M., Stotz, E.H. (eds.). Amsterdam, London, New York: Elsevier 1964
Vallee, B.L., Riordan, J.F., Auld, D.S., Latt, S.A.: Chemical approaches to the mode of action of carboxypeptidase A. Phil. Trans. Roy. Soc. (London), Ser. B 257, 215-230 (1970)
Vallee, B.L., Williams, R.J.P.: Metalloenzymes: the entatic nature of their active sites. Proc. Natl. Acad. Sci. 59, 498-505 (1968a)
Vallee, B.L., Williams, R.J.P.: Enzyme action: views derived from metalloenzyme studies. Chem. Britain 4, 397-402 (1968b)
Vänngård, T.: Some properties of ceruloplasmin copper as studied by ESR spectroscopy. In: Magnetic Resonance in Biological Systems. Ehrenberg, A., Malmström, B.G., Vänngård, T. (eds.). Oxford, London: Pergamon 1967
Vänngård, T.: Copper proteins. In: Biological Applications of Electron Spin Resonance, Chap. 9. Swartz, H.M., Bolton, J.R., Borg, D.C. (eds.). New York, London: Wiley-Interscience 1972

Vault, D. De, Chance, B.: Studies of photosynthesis using a pulsed laser. I. Temperature dependence of cytochrome oxidation rate in *Chromatium*. Evidence for tunneling. Biophys. J. 6, 825-847 (1966)

Vault, D. De, Parkes, J.H., Chance, B.: Electron tunneling in cytochromes. Nature (London) 215, 642-644 (1967)

Vega, J.M., Garrett, R.H., Siegel, L.M.: Siroheme: a prosthetic group of the *Neurospora crassa* assimilatory nitrite reductase. J. Biol. Chem. 250, 7980-7989 (1975)

Vickery, L.E.: Spin States of Hemeproteins by MCD. In: Methods in Enzymology, Vol. Biomembranes, Part C (Biological Oxidations). New York: Academic Press 1978

Vickery, L., Nozawa, T., Sauer, K.: Magnetic circular dichroism studies of myoglobin complexes. Correlations with heme spin state and axial ligation. J. Am. Chem. Soc. 98, 343-350 (1976a)

Vickery, L., Nozawa, T., Sauer, K.: Magnetic circular dichroism studies of low-spin cytochromes. Temperature dependence and effects of axial coordination of the spectra of cytochrome c and cytochrome b_5. J. Am. Chem. Soc. 98, 351-357 (1976b)

Vleck, J.H., van: The theory of electric and magnetic susceptibilities. London, New York: Oxford Univ. Press 1932

Warburg, O.: Heavy metal prosthetic groups and enzyme action. Oxford: Clarendon Press 1949

Ward, K.B., Hendrikson, W.A., Klippenstein, G.L.: Quarternary and tertiary structure of haemerythrin. Nature (London) 257, 818-821 (1975)

Watenpaugh, K.D., Sieker, L.C., Herriott, J.R., Jensen, L.H.: The structure of a non-heme iron protein: rubredoxin at 1.5 Å resolution. Cold Spr. Harbor Symp. Quant. Biol. 36, 359-367 (1972)

Watenpaugh, K.D., Sieker, L.C., Herriott, J.R., Jensen, L.H.: Refinement of the model of a protein: rubredoxin at 1.5 Å resolution. Acta Cryst. B29, 943-956 (1973)

Watson, R.E.: Iron series Hartree-Fock calculations. I. Phys. Rev. 118, 1036-1045 (1960a)

Watson, R.E.: Iron series Hartree-Fock calculations. II. Phys. Rev. 119, 1934-1939 (1960b)

Weinryb, I.: The behaviour of horseradish peroxidase at high hydrogen peroxide concentrations. Biochemistry 5, 2003-2008 (1966)

Weiss, C., Kobayashi, H., Gouterman, M.: Spectra of porphyrins III: self-consistent molecular orbital calculations of porphyrin and related ring systems. J. Mol. Spectr. 16, 415-450 (1965)

Weissbluth, M.: Hemoglobin: Cooperativity and Electronic Properties. Berlin, Heidelberg, New York: Springer 1974

Wentworth, R.A.D.: Mechanisms for the reactions of molybdenum in enzymes. Coord. Chem. Rev. 18, 1-27 (1976)

Weser, U., Bunnenberg, E., Cammack, R., Djerassi, C., Flohé, L., Thomas, G., Voelter, W.: A study on purified bovine erythrocuprein. Biochim. Biophys. Acta 243, 203-213 (1971)

Wherland, S., Gray, H.G.: Metalloprotein electron transport reactions: analysis of reactivity of horse heart cytochrome c with inorganic complexes. Proc. Natl. Acad. Sci. 73, 2950-2954 (1976)

Williams, R.J.P.: The properties of metalloporphyrins. Chem. Reviews 56, 299-328 (1956)

Willick, G.E., Schonbaum, G.R., Kay, C.M.: Circular dichroism and absorption spectra of horse radish peroxidase and sperm whale myoglobin in the Soret region. Biochemistry 8, 3729-3734 (1969)

Wilson, D.F., Dutton, P.L.: The oxidation-reduction potentials of cytochromes a and a_3 in intact rat liver mitochondria. Arch. Biochem. Biophys. 136, 583-584 (1970)

Wilson, D.F., Erecinska, M., Leigh, J.S., Jr., Koppelman, M.: The properties of the mitochondrial succinate-cytochrome c reductase. Arch. Biochem. Biophys. 151, 112-121 (1972)
Wilson, D.F., Leigh, J.S., Jr.: Heme-heme interaction in cytochrome c oxidase *in situ* as measured by EPR spectroscopy. Arch. Biochem. Biophys. 150, 154-163 (1972)
Wilson, D.F., Lindsay, J.G., Brocklehurst, E.S.: Heme-heme interaction in cytochrome oxidase. Biochim. Biophys. Acta 256, 277-286 (1972b)
Wilson, E.W., Jr., Martin, R.B.: Penicillamine deprotonations and interactions with copper ions. Arch. Biochem. Biophys. 142, 445-454 (1971)
Wood, J.M., Brown, D.G.: The chemistry of vitamin B_{12}-enzymes. Struct. Bonding 11, 47-105 (1972)
Wüthrich, K.: Structural studies of hemes and hemoproteins. Struct. Bonding 8, 53-121 (1970)
Wyckoff, H.W., Doscher, M., Tsernoglou, D., Inagami, T., Johnson, L.N., Hardman, K.D., Allewell, N.M., Kelly, D.M., Richards, F.M.: Design of a diffractometer and flow cell system for X-ray analysis of crystalline proteins with applications to the crystal chemistry of ribonuclease-S. J. Mol. Biol. 27, 563-578 (1967)
Yamamoto, H., Saitō, M., Yonetani, T.: Determination of microscopic oxygen equilibrium constants of hemoglobin and cobalt-hemoglobin using cobalt-iron hybrid hemoglobins. Federation Proc. 35, 1392 (1976)
Yamazaki, I., Mason, H.S., Piette, L.: Identification, by electron paramagnetic resonance spectroscopy, of free radicals generated from substrates by peroxidase. J. Biol. Chem. 235, 2444-2449 (1960)
Yonetani, T., Drott, H.R., Leigh, J.S., Jr., Reed, G.H., Waterman, M.R., Asakura, T.: Electromagnetic properties of hemoproteins III. Electron paramagnetic resonance characteristics of iron (III) and manganese (II) protoporphyrins IX and their apohemoprotein complexes in high spin states. J. Biol. Chem. 245, 2998-3003 (1970)
Yonetani, T., Iizuka, T., Asakura, T., Ōtsuka, J., Kotani, M.: Analysis of thermal equilibria between high-spin and low-spin states in mesohemoproteins. J. Biol. Chem. 247, 863-868 (1972)
Zener, C.: Analytic atomic wave functions. Phys. Rev. 36, 51-56 (1930)
Zerner, M., Gouterman, M., Kobayashi, H.: Porphyrins VIII: Extended Hückel calculations on iron complexes. Theoret. Chim. Acta 6, 363-400 (1966)
Zipp, A., Ogunmola, G., Neuman, R.C., Jr., Kauzmann, W.: Effect of pressure on the visible absorption spectrum of metmyoglobin fluoride. J. Am. Chem. Soc. 94, 2541-2542 (1972)

Subject Index

Molecular Biology, Biochemistry and Biophysics

Editors:
A. Kleinzeller
G.F. Springer
H.G. Wittmann

Vol. 1: J.H. van't Hoff
Imagination in Science
Translated into English with notes and a general introduction by G.F. Springer
1 portrait. VI, 18 pages, 1967
ISBN 3-540-03933-3

Vol. 2: K. Freudenberg, A.C. Neish
Constitution and Biosynthesis of Lignin
10 figures. IX, 129 pages. 1968
ISBN 3-540-04274-1

Vol. 3: T. Robinson
The Biochemistry of Alkaloids
37 figures. X, 149 pages. 1968
ISBN 3-540-04275-X

Vol. 5: B. Jirgensons
Optical Acitivity of Proteins and Other Macromolecules
2nd revised and enlarged edition
71 figures. IX, 199 pages. 1973
ISBN 3-540-06340-4

Vol. 6: F. Egami, K. Nakamura
Microbial Ribonucleases
5 figures. IX, 90 pages. 1969
ISBN 3-540-04657-7

Vol. 8: **Protein Sequence Determination**
A Sourcebook of Methods and Techniques
Edited by S.B. Needleman
2nd revised and enlarged edition
80 figures. XVIII, 393 pages. 1975
ISBN 3-540-07256-X

Vol. 9: R. Grubb
The Genetic Markers of Human Immunoglobins
8 figures. XII, 152 pages. 1970
ISBN 3-540-05211-9

Vol. 10: R.J. Lukens
Chemistry of Fungicidal Action
8 figures. XIII, 136 pages. 1971
ISBN 3-540-05405-7

Vol. 11: P. Reeves
The Bacteriocins
9 figures. XI, 142 pages. 1972
ISBN 3-540-05735-8

Vol. 12: T. Ando, M. Yamasaki, K. Suzuki
Protamines
Isolation, Characterization, Structure and Function
24 figures, 17 tables. IX, 114 pages. 1973
ISBN 3-540-06221-1

Vol. 13: P. Jollès, A. Paraf
Chemical and Biological Basis of Adjuvants
24 figures, 41 tables.VIII, 153 pages. 1973
ISBN 3-540-06308-0

Vol. 14: **Micromethods in Molecular Biology**
Edited by V. Neuhoff
275 figures (2 in color), 23 tables
XV, 428 pages. 1973
ISBN 3-540-06319-6

Vol. 15: M. Weissbluth
Hemoglobin
Cooperativity and Electronic Properties
50 figures. VIII, 175 pages. 1974
ISBN 3-540-06582-2

Vol. 16: S. Shulman
Tissue Specificity and Autoimmunity
32 figures. XI, 196 pages. 1974
ISBN 3-540-06563-6

Vol. 17: Y.A. Vinnikov
Sensory Reception
Cytology, Molecular Mechanisms and Evolution
Translated from the Russian by W.L. Gray and B.M. Crook
124 figures (173 speparate illustrations)
IX, 392 pages. 1974
ISBN 3-540-06674-8

Springer-Verlag Berlin Heidelberg New York

Vol. 18: H. Kersten, W. Kersten
Inhibitors of Nucleic Acid Synthesis
Biophysical und Biochemical Aspects
73 figures. IX, 184 pages. 1974
ISBN 3-540-06825-2

Vol. 19: M.B. Mathews
Connective Tissue
Macromolecular Structure and Evolution
31 figures. XII, 318 pages. 1975
ISBN 3-540-07068-0

Vol. 20: M.A. Lauffer
Entropy-Driven Processes in Biology
Polymerization of Tobacco Mosaic Virus Protein and Similar Reactions
90 figures. X, 264 pages. 1975
ISBN 3-540-06933-X

Vol. 21: R.C. Burns, R.W.F. Hardy
Nitrogen Fixation in Bacteria and Higher Plants
27 figures. X, 189 pages. 1975
ISBN 3-540-07192-X

Vol. 22: H.J. Fromm
Initial Rate Enzyme Kinetics
88 figures, 19 tables. X, 321 pages. 1975
ISBN 3-540-07375-2

Vol. 23: M. Luckner, L. Nover, H. Böhm
Secondary Metabolism and Cell Differentiation
52 figures, 6 tables. VI, 130 pages. 1977
ISBN 3-540-08081-3

Vol. 24: **Chemical Relaxation in Molecular Biology**
Editors: I. Pecht, R. Rigler
141 figures, 47 tables. XVI, 418 pages 1977
ISBN 3-540-08173-9

Vol. 25: **Advanced Methods in Protein Sequence Determination**
Editor: S.B. Needleman
96 figures, 25 tables. Approx. 210 pages 1977
ISBN 3-540-08368-5

Structure and Bonding

Editors:
J.D. Dunitz, P. Hemmerich
J.A. Ibers, C.K. Jorgensen
J.B. Neilands, D. Reinen
R.J.P. Williams

Vol. 20: **Biochemistry**
57 figures. IV, 167 pages. 1974
ISBN 3-540-07053-2

A.S. Mildvan, C.M. Grisham: The Role of Divalent Cations in the Mechanism of Enzyme Catalyzed Phosphoryl and Nucleotidyl Transfer Reactions. – H.P.C. Hogencamp, G.N. Sando: The Enzymatic Reduction of Ribonucleotides. – W.T. Oosterhuis: The Electronic State of Iron in Some Natural Iron Compounds: Determination by Mössbauer and ESR Spectroscopy. – A. Trautwein: Mössbauer-Spectroscopy on Heme Proteins.

Vol. 23: **Biochemistry**
50 figures. IV, 193 pages. 1975
ISBN 3-540-07332-9

J.A. Fee: Copper Proteins – Systems Containing the "Blue" Copper Center. – M.F. Dunn: Mechanisms of Zinc Ion Catalysis in Small Molecules and Enzymes. – W. Schneider: Kinetics and Mechanism of Metalloporphyrin Formation. – M. Orchin, D.M. Bollinger: Hydrogen-Deuterium Exchange in Aromatic Compounds.

Vol. 29: **Biochemistry**
51 figures, 48 tables. IV, 219 pages. 1976
ISBN 3-540-07886-X

W.G. Zunft: The Molecular Basis of Biological Dinitrogen Fixation. – J.J.R. Frausto da Silva, R.J.P. Williams: The Uptake of Elements by Biological Systems. – A.M. Cheh, J.B. Neilands: The δ-Aminolevulinate Dehydratases: Molecular and Environmental Properties. – P.J. Sadler: The Biological Chemistry of Gold. A Metallo-Drug and Heavy-Atom Label with Variable Valency.

Springer-Verlag Berlin Heidelberg New York

MIX
Papier aus verantwortungsvollen Quellen
Paper from responsible sources
FSC® C105338

If you have any concerns about our products,
you can contact us on
ProductSafety@springernature.com

In case Publisher is established outside the EU,
the EU authorized representative is:
Springer Nature Customer Service Center GmbH
Europaplatz 3, 69115 Heidelberg, Germany

Printed by Libri Plureos GmbH
in Hamburg, Germany